Epigenetics

Epigenetics

The Ultimate Mystery of Inheritance

Richard C. Francis

W. W. NORTON & COMPANY

NEW YORK   LONDON

For information about permission to reproduce
selections from this book,
write to Permissions, W. W. Norton & Company, Inc.,
500 Fifth Avenue, New York, NY 10110

For information about special discounts for bulk
purchases, please contact W. W. Norton Special Sales
at specialsales@wwnorton.com or 800-233-4830

Manufacturing by Courier Westford
Book design by Lovedog Studio
Production manager: Anna Oler

Library of Congress Cataloging-in-Publication Data

Francis, Richard C., 1953–
Epigenetics : the ultimate mystery
of inheritance / Richard C. Francis. — 1st ed.
    p. cm.
Includes bibliographical references and index.
ISBN 978-0-393-07005-7 (hardcover)
1. Genetic regulation. 2. Epigenesis.
3. Adaptation (Biology) I. Title.
QH450.F73 2011
572.8'65—dc22

                                    2011000696

W. W. Norton & Company, Inc.
500 Fifth Avenue, New York, N.Y. 10110
www.wwnorton.com

W. W. Norton & Company Ltd.
Castle House, 75/76 Wells Street, London W1T 3QT

1 2 3 4 5 6 7 8 9 0

For my mother, Carol Francis,
and my father, Richard W. Francis

Contents

Preface

What Your Genes Are Wearing

HERE'S A PUZZLE. CONSIDER THE CASE OF TWO BROTHERS, when each had reached the age of twenty. One of them—call him Al—was a typical young male. His brother, Bo, however, was not at all typical of young males at this age. Bo looked more like a preadolescent male: poorly developed muscles, absolutely no facial hair, and a voice to match. Their mother was understandably worried about Bo, and soon after his twentieth birthday finally convinced him to see a doctor. Once Bo removed his clothes, the doctor immediately noticed that something was missing—his genitals. A closer inspection revealed that he did in fact have genitals but nothing like those you would expect of a twenty-year-old male. They seemed vestigial. The doctor's diagnosis was Kallmann syndrome, a disorder of sexual development.[1] What's puzzling is this: Al and Bo are identical twins, nature's clones. So what happened to Bo? And why didn't it happen to Al?

Kallmann syndrome is an odd-seeming mixture of developmental defects. Not only is sexual development affected, but so too is the sense of smell. Those who suffer from this disorder have greatly impaired olfaction; some have no sense of smell whatsoever. This strange-seeming association reflects the fact that Kallmann syndrome is a developmental defect in a certain part of the embryonic brain called the olfactory placode.[2] As the name implies, it is from this part of the brain that our olfactory sense develops, but it is also from this part of the brain that certain neurons originate that play a huge role in sexual development. During normal sexual development these neurons migrate from the olfactory placode to the hypothalamus. In those with Kallmann syndrome this migration is disrupted.

It is also noteworthy, therefore, that even though only Bo's sexual development was impaired, both Al and Bo have an impaired sense of smell; both, in fact, have Kallmann syndrome. Why is Bo's case so much more severe? Kallmann syndrome is generally considered a genetic disease.[3] Yet Al and Bo share whatever genetic defects may have contributed to Bo's condition. What is it that they *don't* share? The story of Al and Bo is based on a real case study,[4] one of the more dramatic examples of *discordance* in genetically identical twins. Nature's clones are far from identical, which is why the term "identical twins" has been replaced by *monozygotic twins*.[5] Their discordances sometimes result from essentially random processes at the biochemical level. We are familiar with one form of biochemical randomness, called mutation, which alters the DNA sequence. It is possible but highly unlikely that Bo's DNA mutated after the fertilized egg split, in which case the twins would be genetically different. It is much more likely that the differences in Al and Bo are epigenetic in nature. The term *epigenetic* refers to long-term alterations of DNA that don't involve changes in the DNA sequence itself. Either Al's DNA was epigenetically altered in a way that meliorated his Kallmann syndrome, or Bo's DNA was epigenetically altered in a way that exacerbated it.

The naked gene consists of DNA in the form of the famous double helix. The genes in our cells are rarely naked, however. They are, rather, clothed in a variety of other organic molecules that are chemically attached. What makes these chemical attachments important is that they can alter the behavior of the genes to which they are attached; they can cause genes to be more or less active. What makes these attachments even more important is that they can stay attached for long periods of time, sometimes a lifetime. Epigenetics is the study of how these long-lasting, gene-regulating attachments are emplaced and removed.[6] Sometimes epigenetic attachments and detachments occur more or less at random, like mutations. Often though, epigenetic changes occur in response to our environment, the food we eat, the pollutants to which we are exposed, even our social interactions. Epigenetic processes occur at the interface of our environment and our genes.

Getting back to twins Bo and Al, it is impossible to say whether their differences reflect random or environmentally induced epigenetic differences. Nor can we know, in this particular instance, what genes are involved. It could be the same genes, the mutations of which are implicated in Kallmann syndrome, or the epigenetic differences may occur in altogether different genes that influence sexual development. We need more than one case study to determine these things.

Al and Bo will continue to epigenetically diverge throughout the course of their lives. These epigenetic differences will make Al or Bo more susceptible to Alzheimer's disease, lupus (systemic lupus erythematosus), and cancer, to name a few ailments.[7] The epigenetics of cancer is particularly well studied. In cancer cells, many genes lose their normal methyl attachments—they are *demethylated*. This demethylation results in a host of abnormal gene activities, one consequence of which is unbridled cell proliferation. It is this global demethylation, not any particular mutation, which is the hallmark of cancer. This is good news, because unlike mutations epigenetic changes are reversible. The goal of much medical epigenetics is to find ways to reverse pathological epigenetic events. Many see in epigenetics the potential for a medical revolution.

Another active area of epigenetic research concerns the fetal environment. Al and Bo are less different epigenetically than non-twin brothers because they shared similar environments throughout their lives. This is especially true of the environment they experienced in the womb. Whatever their mother's diet during that period, it affected them equally. The same goes for whatever stress she experienced during pregnancy. More typical siblings, however, can experience quite different fetal environments. The epigenetic alterations that result will make one or the other more susceptible to obesity, diabetes, heart disease, and atherosclerosis, as well as depression, anxiety, and schizophrenia.

Though the epigenetics of what ails us is the most topical, other sorts of epigenetic processes are, for a biologist, more fundamental. Particularly important is the problem of development: how a fertilized egg can become you or me. The problem of development can be broken out into subproblems. There have been major advances in solving one of these subproblems, called cellular differentiation, because of epigenetic research. We all passed through a stage in which we were a hollow ball of generic cells, called *stem cells*. These stem cells are not only genetically identical; they are physically indistinguishable as well. How, then, do we come to have skin cells, blood cells, neurons, muscle cells, bone cells, and so forth, all of which remain genetically identical? Epigenetics holds the key to unlocking this mystery.

Epigenetics also informs some mysteries concerning inheritance. Our parents make separate but equal genetic contributions to who we are. They also make separate but *unequal* epigenetic contributions. For some genes it makes a difference whether you inherit them from your father or your mother. These genes are epigenetically activated if they come by way of your mother, but inactivated if they come by way of your father (and vice versa). Other epigenetic states, some environmentally induced, can be transmitted from grandparent to grandchild.

This book is intended as an introduction to epigenetics for those unfamiliar with this exciting new field of research. It is written for the nonspecialist who seeks to be informed about this important subject. The scope of epigenetics is too vast for a comprehensive treatment, which would be inappropriate for the intended audience in any case. I will cover only some of the highlights and hope thereby to impart a sense of what's going on.

I have a secondary agenda as well, which concerns the implications of epigenetics. I believe that epigenetics should substantially alter the way we think about genes, what they are and what they do, particularly what they do with respect to our development from a fertilized egg. In the traditional view, genes function as executives that direct the course of our development. In the alternative view, which I advocate, the executive function resides at the cellular level and genes function more like cellular resources. I have tried to present the material in this book so that those uninterested in my secondary agenda will nonetheless learn something of value about epigenetics.

Throughout this book I emphasize research that relates most

directly to the human condition, primarily because I believe this is the best way to connect with nonscientists. Humans don't make great research subjects though, for both ethical and practical reasons. Some of the best epigenetic research is on plants, but I only refer to that work when I cannot find better examples closer to home. I focus, rather, on animal models, and especially mammals. I do not emphasize in the text—as is the norm for many popular science books—specific labs, researchers, or experiments. This is in large part to help the narrative flow. I cover too much ground and the work of too many researchers to make one or a few central to my project. Rather, I want to keep the reader focused on what the research has revealed. The interested reader who wants more information on the researchers and research covered will find it in the Notes.

I have diligently strived to keep the main text as nontechnical as possible. For those interested in more detail, again look to the Notes for each chapter. The epigenetic topics I discuss are covered in a particular order, such that each chapter builds to some extent on the preceding chapters.

Chapter 1 concerns a historical event, the Dutch famine of World War II and its epigenetic consequences. During the course of the next several chapters the reader is provided the tools, in a stepwise manner, for understanding how it is that a famine could influence the long-term health not only of those who experienced it in the womb but of their children as well. First, in Chapter 2, I supply some basic background about genetics that will be essential for understanding epigenetics, including the crucial concept of gene regulation. Chapter 3 concerns garden-variety gene regulation, that is, what was known about gene regulation before we came to understand epigenetic gene regulation. Chapters 4, 5, and 6 delve into epigenetic gene regulation and how it is influenced by the environment, beginning in the womb. In Chapter 7 we turn to the inheritance of epigenetic states, including those induced by the fetal and social environment. At this point we can better understand why the effects of the Dutch famine persist to this day. In the remainder of the book, we move beyond anything that can be gleaned about epigenetics from the Dutch famine example, to explore what are, for biologists, the most significant applications of epigenetics, including stem cells and cancer.

Epigenetics

Chapter 1

A Grandmother Effect

ONE OF THE LESSER-KNOWN ATROCITIES COMMITTED DURING World War II occurred during its waning months. In September 1944, the Germans were in retreat throughout most of the Europe. They retained, however, a stronghold in the populous northwestern portion of the Netherlands, which was of both strategic and symbolic importance to the fading Nazi cause. But German control of this area was threatened by Allied forces approaching from the south, in support of which the exiled Dutch government ordered a railway strike. Though the allied forces were stopped at Arnhem, the Germans retaliated for the railway strike and other hostile actions by Dutch partisans with a food embargo. Unfortunately, the embargo coincided with the onset of a particularly severe winter during which the canals froze over, disrupting barge transport. Things further deteriorated when, in response to the advance of allied troops from the south, the retreating Germans destroyed what remained of the transportation infrastructure and flooded much of western Holland's agricultural lands.

By the end of November, the diet for most inhabitants of the major cities in western Holland, including Amsterdam, was reduced to only 1,000 calories per day, a huge drop from the 2,300 calories normally consumed by an active woman and the 2,900 calories normally consumed by an active man.[1] At the end of February 1945, rations had dropped to 580 calories in some parts of western Holland. To augment this meager fare—consisting largely of bread, potatoes, and a cube of sugar—city dwellers were forced to walk many miles to the nearest farms, where they traded whatever they owned for food. Those without the means to trade were forced to eat tulip bulbs and sugar beets as a last resort. The worst effects of the famine were largely confined to the major cities of western Holland, particularly the poor and middle class. In the rural areas of the west, farmers were self-sustaining. Eastern Holland—roughly half of the Dutch population—largely escaped the famine.

By the time the Netherlands was liberated by the Allies in May 1945, 22,000 people had died in western Holland. Death by starvation is the traditional measure of a famine's effects. But that measure, it turns out, is inadequate, for many who survived the famine were also severely affected, not least those who experienced the famine in their mothers' wombs. This group became part of the Dutch Famine Birth Cohort Study, a pioneering investigation of malnutrition that continues to this day.[2]

The Dutch famine was unique in that its onset and end could be precisely dated. Moreover, the Dutch maintained and stored meticulous health records for all citizens after this period. These two circumstances comprised what scientists refer to as a *natural experiment*. Clement Smith was the first person to recognize it as such.[3] Smith, of the Harvard Medical School, was among a group of doctors from the United Kingdom and the United States who were flown into the Netherlands in May 1945, immediately after the German surrender. He saw in this tragedy an opportunity to advance our understanding of the effect of maternal nutrition on fetal development.

Some Unexpected Consequences

Smith obtained obstetric records from The Hague and Rotterdam. He found that babies born during the famine weighed considerably less than those born prior to the famine. That this does not seem surprising to us now is due in no small part to Smith's groundbreaking research efforts. Moreover, as Smith suspected, subsequent research established a strong link between low birth weight and poor neonatal health.

Others wondered about the longer-term effects of the famine. The first long-term effect was identified, retrospectively, in eighteen-year-old military conscripts. Those who were in their mother's womb during the famine came of age for military service—which was compulsory for males—in the early 1960s. At induction they were given a thorough physical examination. These records were subsequently inspected by a group of scientists in the 1970s.[4] They found that those exposed to the famine during the second and third trimester evidenced significantly elevated levels of obesity, roughly double the levels of those born before or after the famine.

A subsequent study, which included both males and females, focused on psychiatric outcomes. Here again the Dutch penchant for detailed medical records made the study possible. The investigators who mined these data found a significant increase in the risk for schizophrenia in those prenatally exposed to the Dutch famine.[5] There was also evidence of an increase in affective disorders, such as depression. Among males, there was an increase in antisocial personality disorder.

In the early 1990s, a new series of studies commenced, based on individuals identified at birth from hospital records, most notably, Wilhelmina Gasthuis Hospital in Amsterdam. The first of these studies was restricted to females and focused primarily on birth weight. The investigators again found that those exposed to the famine during the third trimester were abnormally small at birth. But they also found that those exposed during the first trimester were larger than average, suggesting some compensatory response, perhaps in the placenta, to food stress early in pregnancy.[6]

In the second study of this series, which commenced when the cohort had reached 50 years of age, both males and females were included. For the first time, investigators turned their attention to cardiovascular and other physiological functions. At this age, those prenatally exposed to the famine were more prone to obesity than those not exposed. Moreover, they showed a higher incidence of high blood pressure, coronary heart disease, and type II diabetes. When the cohort was resurveyed at the age of fifty-eight years, these health measures continued to trend adversely.[7]

But the nature of the adverse effects of the famine on the fetus depended largely on the timing of exposure. For instance, coronary heart disease and obesity were associated with early exposure during the first trimester. Women exposed during the first trimester also had an increased risk of breast cancer. Those exposed during the second trimester had more lung and kidney problems. Altered glucose intolerance was most evident in those exposed during late gestation.[8]

By the late 1990s, several research groups were independently studying the Dutch famine cohort, studies which continue to this day. Together they provide some of the most compelling evidence for the long-term effects of the fetal environment on our health. Having documented these effects of the famine, some of the scientists involved have turned their attention to the mechanism underlying them. That is, they now seek to understand how mothers' malnutrition during pregnancy can cause obesity or schizophrenia in their offspring when those offspring are adults.

From Environment to Gene

It will come as a surprise to many that our external environment affects us through our genes, by modulating their activity. Our environment does not affect our genes directly. Rather environmental influences on our genes are mediated by changes in the cells in which our genes reside. Different kinds of cells respond differently to the same environmental factor, whether it is social stress or food deprivation in the womb. As such, and despite the fact that all of the cells in our body have the same genes, any environmental effect in you is cell type–specific. Your liver cells will react one way to poor nutrition, your neurons will react in a different way, and many cell types won't react at all. Therefore, in determining any environmental influence on gene action, scientists look at specific cell populations, such as neurons in a particular part of the brain, liver cells, pancreatic cells, and such.

The Dutch famine clearly affected many different kinds of cells in the exposed individuals, some in the brain, some in the heart, some in the liver, some in the pancreas, and so forth. If we were to compare, say, the liver cells of those in the Dutch famine cohort with those unaffected by the famine, we are likely to find different patterns of gene activity. Some genes in the liver cells of affected individuals will be more active and some less active than in unaffected individuals. The initial goal is to identify the particular genes in these liver cells that are altered activity-wise by food deprivation in the womb. Then comes the hard work of establishing a causal link between these altered gene activities in the liver cells and diabetes or whatever condition we seek to explain.

The control of gene activity by a cell is called *gene regulation*. I will discuss gene regulation, especially epigenetic gene regulation, in more detail later in the book. For now, I am painting with a broader brush.

Before the advent of epigenetics, biologists already knew a great deal about short-term gene regulation, that is, gene regulation that occurs over time spans ranging from minutes to weeks. I will refer to this short-term gene regulation as "garden-variety" gene regulation, because this is the form of gene regulation long taught in introductory biology courses. Epigenetic gene regulation is not garden-variety gene regulation. For reasons we will explore later, epigenetic gene regulation occurs over much longer intervals, sometimes spanning an entire lifetime. Epigenetic gene regulation is long-term gene regulation. It is the kind of gene regulation that is most relevant to the Dutch famine cohort.

Epigenetically regulated genes can be identified by characteristic marks in the form of particular chemical attachments. The most common sort of chemical attachment involves the methyl group, which consists of one carbon atom bonded to three hydrogen atoms (CH_3). A gene with methyl attachments is said to be *methylated*. Methylation is not an all-or-none affair; genes can be methylated to varying degrees. Generally, the more methylated a gene is, the less active it is. It is with these facts in mind that scientists have begun to look for epigenetic alterations induced by the Dutch famine. Though these are still the early days, this research has already borne fruit.

In one recent study of the Dutch famine cohort, a number of epigenetically altered genes were identified in blood cells.[9] That is, the degree of methylation in these genes differed in those exposed to the famine compared with those who were not exposed. Of particular note were the epigenetic differences in a gene that codes for the hormone *insulin-like growth factor 2* (IGF2), so called because it closely resembles insulin and because it promotes growth, through cell division, in a variety of cell types. (The "2" reflects the fact that it was the second of three IGF molecules to be discovered.) IGF2 is essentially a growth hormone, one that is particularly important for the growth of the fetus.

Scientists are far from being able to causally connect the epigenetic alteration in *IGF2*, the gene for IGF2, to any of the Dutch famine's diverse health impacts, such as birth weight, diabetes, and schizophrenia. For starters, they will need to determine whether similar epigenetic changes in *IGF2* can be found in other types of cells. They will then need to establish a causal link between the cell type–specific epigenetic alterations in *IGF2* and these conditions. This result is nonetheless quite significant in demonstrating that the epigenetic effects of the fetal environment can extend over six decades.

Most epigenetic attachments are removed during the production of sperm cells and egg cells. Hence, the fertilized egg commences development with an epigenetically clean slate. Sometimes, though, epigenetic attachments can be passed on, along with the genes to which they are attached, to the next generation. It is noteworthy, in this regard, that the adverse effects of the famine were not confined to those who lived through it. The children of those who experienced the famine through their mother's womb are more prone to ill health later in their lives than children of mothers not exposed to the famine.[10]

This is really quite an astounding discovery, a nongenetic mode of inheritance that influences our health. As I will discuss later in the book, scientists are increasingly aware of nongenetic inheritance of various sorts, some of which we can call true epigenetic inheritance. It is far from clear, however, that this grandmother effect of the Dutch famine represents true epigenetic inheritance, that is, the inheritance of methylated genes. As we will see, there are other possible explanations. To better understand whether this grandmother effect is or isn't true epigenetic inheritance, we need some background. I begin with the stuff to which epigenetic marks are attached: What, exactly, are these things we call genes? And what do they actually do?

Chapter 2

Directors, Actors, Stagehands

IT WASN'T A TYPICAL BIOLOGY LAB, NOT BY TODAY'S STAN-
dards, not by the standards of the time—1910. The first thing a
visitor would notice, long before he entered, was the smell. That
you could smell the lab wasn't so unusual; many biology labs ema-
nate odd odors of various sorts. But this smell wasn't like any of the
typical biological lab smells; it was distinctly funky, like the smell
of a metal garbage bin baking in the sun behind a supermarket.

Visually the lab was no less unprepossessing; it was small and
filthy. An impressive layer of detritus had accumulated on the
floor, home to a flourishing population of cockroaches. This lab
was as notable for what it lacked as for what was uncharacteristi-
cally present: no flasks or beakers or test tubes or pipettes. The only
observable glassware was used milk bottles scattered haphazardly
throughout. Also missing were microscopes, even the simplest low-
power kind. Functioning in their stead was an assortment of hand
lenses, like those that the elderly used to use for reading before the
advent of reading glasses.

Also lacking was any sense of formality or hierarchy. This seemed particularly odd to visitors from European universities, most of which were modeled after those in Germany. There, the most informal way of addressing the person who headed the lab was "Herr Doktor Professor." Moreover, a German professor's office was always closed and he was available only by appointment. Here, the door to the professor's office, located at one end of the lab, was always open, and those working in the lab approached it seemingly at the merest whim and without deference. Moreover, they addressed the professor by his first name, a practice that is now commonplace in American universities but was not then, and certainly not elsewhere in the world.

Yet in this humble lab, the infant science of genetics was nurtured to a degree unrivaled in the rest of the world. On any given day you could find at their labors two future Nobel prize laureates, as well as several other scientists who were to shape the course of genetics. First among them was the man who occupied the lone office, Thomas Hunt Morgan, whose significance in the history of genetics is second only to that of the Moravian monk Gregor Mendel.[1] Morgan's goal was to determine the location of Mendel's "hereditary factors"—now called *genes*—on particular chromosomes. Morgan's gene mapping was much different than today's gene mapping. The technology was not then available to directly locate genes on chromosomes. Instead, he had to take a much more indirect route. He could only identify a gene through a mutation that caused some observable change in the appearance (phenotype) of his subjects. If this mutation was correlated with a different trait, he could assume that the genes for the two traits were located on the same chromosome. The more highly correlated they were, the closer together they must be.

Morgan was a scion of southern gentry whose immersion in

science had helped him successfully navigate a culturally jarring transition to New York City. His lab at Columbia University was on the uppermost floor of Schermerhorn Hall, up-atmosphere from the more traditional biology labs and their more traditional odors. Morgan's lab was called the Fly Room by its inhabitants, not because of the numerous houseflies that competed with the roaches for the lab's effluvium, but for the much smaller critters that occupied all of those empty milk bottles: fruit flies. Though the fruit flies themselves don't stink, they were responsible for the lab's odors and dishevelment. For fruit flies, as their name suggests, eat fruit; they deposit their eggs in fruit as well. They prefer their fruit overripe—rotten by human standards—and to keep them happy there was plenty of rotting fruit, especially bananas, strewn about the lab. In fact, according to lab lore, it was someone's lunch banana left inadvertently on a window sill that initially attracted fruit flies to the lab.

The utility of fruit flies was not immediately apparent to Morgan, though. Originally he wanted to use mice for his experiments. But mice had their limitations, given Morgan's goal. Morgan needed subjects that have short life spans and can produce numerous generations per year. As mammals go, mice are pretty good in that respect, but all mammals, including mice, are relatively slow breeders compared with insects and many other invertebrate animals. So it was fortunate, in retrospect, that Morgan's early mouse-based grant applications were turned down and he was forced to "go outside the mammal box"—way outside. He settled, fatefully, on fruit flies, which were readily available, easy to maintain in the milk bottles, and could produce in one year a whopping fifty generations. Though he couldn't know it at the time, fruit flies would remain the animal of choice for many genetic investigations to this day.

Initially, though, it wasn't at all obvious that Morgan had made the right choice. Notwithstanding the many generations of fruit flies bred in the lab, after two full years nary a mutation could be found. Morgan was close to despair for having wasted so much valuable research money and time on this increasingly quixotic-seeming quest. It wasn't that the fruit flies weren't mutating: all living things experience mutation; it is a fact of life. But the primary virtue of fruit flies—their short generations—came at a price: small size. As such, the only fruit fly mutations that the Fly Room scientists could identify would be those that caused some gross alteration in a fruit fly's appearance—thus could be spotted through one of those hand lenses—yet was not lethal. Those mutations are extremely rare.

Finally, though, in the third year of their quest came their first success—a fruit fly born with white eyes. Normal fruit flies have wine-red eyes. Morgan referred to the red-eyed flies as the *wild type*. The white-eyed mutants were in fact blind, but they could still breed under the right conditions. After numerous crosses between white-eyed mutants and wild-type red-eyed individuals, Morgan and his coworkers were able to trace the white-eye mutation to a particular chromosome—a sex chromosome, as it happens.

At this point, some important terminological distinctions are in order. Let's start with the notion of a *chromosome*. Chromosomes ("colored bodies") are so called because they look purplish-brown under a microscope. By Morgan's time, most scientists were convinced that genes resided on chromosomes. Though they did not know what chromosomes physically consisted of, chromosomes were assumed to be linear. One common conception was that genes were like beads on a chromosomal string. That simile will do for now. A gene, then, has a particular location on a chromosome, an address called a *locus* (plural *loci*). Sometimes there is only one genetic variant at a locus, but more often two or more. These variants are called *alleles*. Think of alleles as the different colors of beads found at a locus. Some loci have beads of only one color (that is, one type of allele), whereas most loci have two or more types of alleles and hence two or more colors.

Morgan found a mutation in a gene that affects eye development. This mutation caused the formation of a new allele, a differently colored bead, at that locus. It was this new allele that Morgan labeled "white." There is no white-eye locus, only a white-eye allele at a locus that influences eye development. We humans have a couple of gene loci that influence eye color, one of which has two variant alleles that largely determine whether our eyes are brown or blue.

There are actually two distinct senses of the term *allele* in genetics. In the foregoing, I have treated alleles as types. But there is another sense of allele—as tokens. By token, I mean the particular instance of a certain type. We inherit two allele tokens at each locus, one from each parent. If the two allele tokens are of the same type, we are *homozygous* for that locus. If the two allele tokens are of different types, we are *heterozygous* for that locus. Let's consider the trait of human eye color. I will assume, for the sake of exposition, that just one locus and two allele types are involved in the determination of eye color.

If we are homozygous for the "brown" allele, our eyes will be brown; if we are homozygous for the "blue" allele, our eyes will be blue. If we are heterozygous, however, things are more complicated. The two alleles could have equal influence, in which case our eye color would be intermediate (some shade of green). Often, though, one allele type has a greater effect on a trait than the other; sometimes the "stronger" allele completely masks the "weaker" allele in the heterozygous condition. In those cases in which there are pronounced differences between stronger and weaker alleles, the stronger allele is said to be *dominant* and the weaker allele *recessive*. For human eye color, the brown allele tends to be dominant and the blue allele recessive. The convention is to label the dominant allele with an uppercase letter and the recessive allele with a lowercase letter. The brown allele would be denoted with a "*B*," the blue allele with a "*b*." That is all the classical genetics you need to know for the purposes of this book.

The Gene Incarnate

Morgan, following Mendel, defined genes with respect to traits, such as eye color. He assumed, again following Mendel, that one gene (locus) corresponded to one trait. And that the gene variants (alleles) corresponded, in some straightforward way, to the variant traits: red or white eyes in fruit flies, brown or blue eyes in humans. That was a reasonable place to start. But it soon became apparent to all but a few unreconstructed Mendelians that most traits weren't like eye color, but were more like height. That is, most traits vary quantitatively (continuously), not qualitatively (discretely). Moreover, differences in height must result from the contributions of many genes, as well as a host of environmental factors.

Morgan himself was not at all interested in what genes were physically, that is, with the material nature of genehood. For his purposes it was sufficient to know that they were units of inheritance that resided on chromosomes, and that they often came in more than one flavor. It was left to others to discover the physical (biochemical) gene.

The first step in this quest was to figure out what chromosomes consisted of. Chromosomes, it turned out, were comprised of two distinct kinds of biochemicals: DNA and proteins. The question then became whether it was the DNA or the proteins that acted as the hereditary material. The issue was decided in favor of DNA through a series of groundbreaking experiments. But now there was a new problem. Though proteins could not be the hereditary material, they—and not DNA—were clearly the primary physiological actors within cells. Some proteins are enzymes that catalyze biochemical reactions, others serve to bind and transport essential elements and chemicals, and still others comprise the structural elements in muscles, skin, and cartilage. Somehow, these essential proteins must be made from DNA. But proteins come in multitudinous varieties while all DNA seemed similar. How then, were all of these proteins made from seemingly unvarying DNA? To answer this question, scientists had to take a closer look at DNA.

It was discovered that the DNA molecule generally has two strands that coil in a helical fashion, the double helix. The "D" in DNA stands for the sugar deoxyribose (NA = nucleic acid). The deoxyribose sugar groups, each separated by a phosphate molecule, comprise the backbone of the DNA molecule. Attached to each sugar is a chemical called a *base* (as in basic, as opposed to acidic). The bases come in four varieties—adenine, cytosine, guanine, and thymine—which are generally denoted by their first letter: A, C, G, and T, respectively. The base from each strand is bound to a base on the other, connecting the two strands like steps on a ladder. But A can only bind to T (and vice versa), and C to G (and vice versa). So there are four types of steps: A-T, T-A, C-G, and G-C.

Francis Crick and James Watson are celebrated for their characterization of DNA as just described, and for their suggestion that the base sequence might relate to the composition of proteins.[2] Soon thereafter came the discovery of the genetic code. This code concerns the mapping of the DNA base sequence to amino acids,

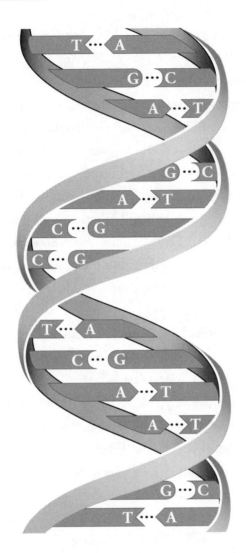

A diagram of the DNA molecule, showing double helix
and base pairing. Fig. 5.28 (p. 85) in N. A. Campbell, *Biology*,
4th ed., 1996.

which are the building blocks of proteins. This mapping is not
nearly as precise as human-devised codes, such as Morse code.

The genetic code implied that genes must consist of linear base
sequences. But where are the boundaries? How can you tell where
one gene stops and another begins? By the late 1950s these questions

seemed answered. Morgan's "one gene (locus) = one trait" became "one gene (locus) = one protein."[3] This formulation provided a straightforward way to delineate genes on a chromosome. Geneticists simply had to find the place on the chromosome where coding started and stopped. But this wonderfully simple formula of "one gene = one protein" was, it turned out, too simple. The relationship between genes and proteins is not nearly that straightforward.

What Genes Actually Do

The process whereby proteins are constructed from genes is called protein synthesis. *Protein synthesis* is a two-stage process. During the first stage, called transcription, one strand of the double helix serves as a template for the creation of a molecule called messenger RNA (mRNA). The term *transcription* is meant to connote the transfer of information from one medium to another, as in musical transcription from, say, piano to guitar. In this case, the transcription is from DNA to RNA.

During the second stage, called translation, the mRNA serves as a template for the creation of a protoprotein. The term *translation* is meant to connote a larger transformation of this information, like that which occurs when one language is translated into another. In protein synthesis, the translation is from the language of the base sequence of the RNA to the amino acid sequence of the protoprotein. The protoprotein is generally not functional. It must be further transformed into a functional protein through a process called posttranslational processing. Posttranslational processing can render the functional protein quite different than what would be predicted given the original DNA sequence alone.[4]

It is tempting to view a gene as the instructor of protein synthe-

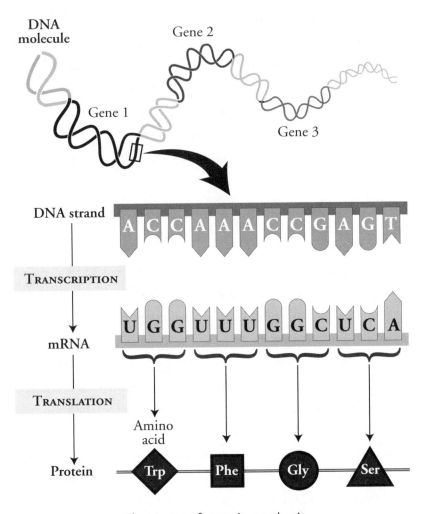

The stages of protein synthesis.

Fig. 16.4 (p. 302) in N. A. Campbell, *Biology*, 4th ed., 1996.

Note: Thymine (T) in DNA replaced by Uracil (U) in mRNA.

sis, to attribute to the gene an executive function. Here is a meta-phor that may illuminate this notion of the executive gene. Think of the cell as a theatrical production, a play. On this view, the gene functions as the director of the play, the proteins are the actors, and all of the other biochemicals in the cell function as stagehands. Genes direct the construction of proteins, through which they con-

trol cellular activities, including the construction of all other biochemicals (such as lipids and carbohydrates), which, in their turn,
labor to the genes' ends.

The problem with this view is that it gives the gene too much
credit for what goes on during protein synthesis, and in the cell
generally. The role of the gene in protein synthesis is to serve as an
indirect template for a protoprotein. This templating function is
crucial but does not render a gene an executive, any more than the
particular plate from which this page was printed functioned as an
executive during the printing process.

An alternative to the executive gene is what I will refer to as the
"executive cell." From this perspective, genes look more like members of an ensemble cast of biochemicals, the interactions of which
constitute a cell. The executive function resides at the cellular level;
it cannot be localized in its parts.[5] Genes function as material
resources for the cell. On this view, each stage of protein synthesis
is guided at the cellular level.[6] But most fundamentally, the "deci-
sions" as to which genes will engage in protein synthesis at any point
in time is a function of the cell, not the genes themselves. That is,
gene regulation is a cellular activity. This is true for both garden-
variety gene regulation and epigenetic gene regulation. Epigenetics
is one form of cellular control over gene activity on this view.

Genes and Traits

Genes influence our traits through the proteins constructed from
them. The eye color locus of fruit flies codes for a protein that
transports red and brown pigments across cell membranes. Morgan's mutant white-eye allele has a different base sequence than the
wild-type allele, and hence codes for a protein that is deficient in

that respect. As a result, the fly ends up with eyes that have no pigment and thus appear white. A defect of a similar transport gene in humans often results in cystic fibrosis. Much work on the genetics of human diseases follows a similar script: a mutant allele causes a specific defect in development. Some of the more topical of such defects are obesity, diabetes, breast cancer, depression, schizophrenia, and substance abuse. And invariably, one or more such mutant alleles have been discovered. Hence the talk of genes for obesity, breast cancer, schizophrenia, substance abuse, etc. It is important to note, however, that the mutant alleles discovered to date account for only a small fraction of these disease states.[7]

Prior to the advent of epigenetics, this quest for mutant alleles (altered base sequences) dominated biological research on these diseases. But disease researchers have become quite cognizant of epigenetics of late, hence the mutant allele approach is now augmented with a search for mutant *epialleles*, that is, alleles with abnormal epigenetic marks.[8]

There is another, quite different sort of explanation in which genes figure, which concerns the normal course of development. With respect to Morgan's fruit flies, the issue would be: how do they usually come to have red eyes? For a fly to have red eyes, it must first of course have eyes. To have eyes, it must first have a nervous system, and so on. In explaining these increasingly generic features of normal development, biologists typically zoom out a bit and shift from the executive gene to the "executive genome." In essence, on the traditional view, normal development results from the coordinated activities of executive genes, which collectively comprise a genetic program.

On the executive cell account, development is a function of coordinated gene actions as well. But this coordination is not programmed into the DNA sequence; rather, it emerges as a result of the cells' interactions with their environment, most notably other cells. We will explore these two views later in the book; for now, it is sufficient to note that epigenetic processes are at the heart of both views.

The Elusive Gene

There used to be a consensus as to what constitutes a gene, but no longer.[9] In the 1960s, there was one gene concept, embodied in the "one gene = one protein" rule. I will call this "the canonical gene." The canonical gene was soon stretched, but not unduly, by the discovery that many genes code for more than one protein. More recent developments, though, have stretched the gene concept beyond recognition. Now pieces of DNA are called genes that don't code for proteins at all.[10]

For the purposes of this book, a gene has two components: a protein-coding sequence and a control panel.[11] The latter is a regulatory region to which proteins and other chemicals bind, either inhibiting or promoting transcription. This is where all garden-variety gene regulation and some epigenetic gene regulation occur. The control panel is not usually considered part of the gene proper because it is not transcribed. But the control panel and the coding sequence comprise a functional unit, so the two will be combined under the gene rubric here.

Much, perhaps most epigenetic gene regulation occurs via attachments outside of the gene proper, even on this expanded definition of genehood. That is, the chemical attachments occur outside of the gene that is epigenetically regulated. In fact, epigenetic attachments can affect genes quite distant from the point of attachment. So it is best to think of epigenetic processes as modifications of DNA, not just of individual genes.

We can think of epigenetics as a new way of looking at DNA that goes beyond the base sequence. The linear base sequence comprises one, albeit the primary dimension of the physical gene, but DNA is a three-dimensional molecule. Epigenetics is a science that extends the study of genes from one to three dimensions. These extra dimensions are particularly important for understanding gene regulation, which is where the epigenetic action is. First, though, let's consider garden-variety gene regulation.

Chapter 3

What Roids Wrought

WHEN JOSÉ CANSECO WAS ARRESTED AT THE MEXICAN BORDER for possession of a female fertility drug, baseball's steroid scandal took a curious but predictable turn. It was Canseco who, after years of rumors, cast the brightest light on the prevalence of anabolic steroids in major league baseball, in his book, *Juiced: Wild Times, Rampant 'Roids, Smash Hits, and How Baseball Got Big*. Initially, though, the book was roundly booed—dismissed as the ravings of a vengeful malcontent who had squandered his talent. There was much truth in this assessment of Canseco the man. In his third year he became the first player ever to hit forty or more home runs and steal forty or more bases in one season. But his decline from that high point was precipitous, as he increasingly focused on swinging for the seats at the expense of his batting average and especially his (outfield) defensive skills. Within a few years of his historic feat, he was better known for having a ball bounce off his head for a home run. Nor did his seemingly blasé attitude endear him to fans or teammates. His first manager, Tony La Russa, then of the Oakland Athletics, came to particularly despise José. In an act of unprecedented spite, La Russa notified Canseco (through a second party) that he had been traded to the Texas Rangers while Canseco stood in the On Deck circle waiting to bat. It was telling that Canseco received little sympathy for his humiliation.

So the public, understandably, was not ready to take Canseco at his word. Soon after the publication of *Juiced*, however, its general veracity became apparent. Canseco himself admitted to being a long-time user, which surprised nobody. But he outraged the baseball community by implicating others by name, many of them All Stars, including his teammate Mark McGwire. By the time the steroid scandal resulted in a congressional investigation, Canseco's seemingly wild claims were largely substantiated, which, of course, became the subject of his second book, *Vindicated: Big Names, Big Liars, and the Battle to Save Baseball.*

Anabolic steroids are popular with baseball players for the same reason they have long been popular in track and field (especially among sprinters), weight lifting, body building, and a host of other athletic endeavors: given the proper training regimen, they promote muscle growth. That is what "anabolic" in "anabolic steroid" refers to. All anabolic steroids are synthetic forms of androgens, particularly the hormone testosterone; that is, they are designed to mimic testosterone for the purpose of muscle building. In this they have a well-documented efficacy. But testosterone has many other effects aside from muscle building. In males, naturally produced testosterone promotes genital development, hair growth, and acne. It also has a host of effects in the brain that affect behavior, most notably libido, but also mood and aggression. These are all considered side effects by steroid users; they are actually just unwanted (by baseball players) but natural responses to this hormone. Many of these side effects are associated with adolescence, when testosterone levels naturally surge. In fact, in many ways, adult steroid users experience perpetual adolescence.

When testosterone levels are unnaturally elevated with synthetic steroids, some side effects become especially problematic. For example, a number of acts of violence have been attributed to so-called roid rage. But many of the undesirable effects of synthetic steroids are caused by their effects on natural testosterone. The body adjusts to the unnaturally high levels of testosterone by shutting down its own production of testosterone. Since synthetic testosterone can only be taken for a few weeks at a time without catastrophic consequences, testosterone levels are unnaturally low during the fallow periods, resulting in depression and low libido. Further complicating matters for users is the fact that one of the byproducts of testosterone metabolism is the estrogen estradiol. One of the consequences of the elevated levels of estradiol is the development of female-like breasts, commonly referred to as "man boobs." High levels of estradiol and low levels of natural testosterone together are responsible for what is, in the macho world of athletics, one of the most undesirable consequences of steroid abuse: shrunken testicles. Even worse, though their libido remains elevated, many longtime users experience erectile malfunctions, a truly ironic case of "the spirit is willing but the flesh is weak."

Which brings us back to Canseco's border bust. Canseco was not only a snitch but a proudly self-proclaimed user and advocate of synthetic testosterone. He maintained his regular steroid regimen after his baseball career was over because he liked the way it made him look and feel. As such a longtime user, Canseco was especially vulnerable to shrinkage and limpness, not only when he had cycled off steroids, but as a more or less permanent state of affairs. At the time of his arrest, his sperm production had probably virtually shut down, hence the fertility drug.

What Canseco had in his possession was another hormone, called *chorionic gonadotropin*, which was purified from gallons of urine garnered from pregnant women. Gonadotropins stimulate the gonads to do what gonads do. What the gonads do is, of course, different in men and women. In women, gonadotropins stimulate the development of eggs in the ovary and the production of estrogens. In men, gonadotropins stimulate the production of sperm and androgens in the testes. That Canseco had pregnant-female-derived gonadotropins is simply because most pharmaceutical gonadotropins come from pregnant women. That he didn't have a prescription is perhaps understandable.

Obviously, androgens are potent chemicals. In this chapter we will explore the reasons for this potency, especially their role in garden-variety gene regulation. This exploration of short-term gene regulation will provide some useful background for the discussion of epigenetic gene regulation in subsequent chapters.

Same Genes, Different Effects

Most of the time, most of your genes in most of your cells are silent; they just sit there, doing nothing. These silent genes must be activated in order to participate in protein production. Activation occurs when certain kinds of chemicals bind with their control panels, initiating the process of transcription as described in the previous chapter. These chemicals are called *transcription factors*. Sex steroids (androgens and estrogens) are important transcription factors. When Canseco is on his steroid cycle, the genes for which testosterone acts as a transcription factor are much more active than they are when Canseco is off his regimen. If we could equate gene activity with light levels, these androgen-sensitive genes would glow much brighter when Canseco was taking steroids.

But only in certain cells. In most of Canseco's cells, the glow from these androgen-sensitive genes would remain dim, even when he was taking the steroids. Yet these steroids circulate widely in the blood, so in principle at least, every cell in his body would be exposed to the steroids. And every cell in Canseco's body has the same genes. So why would only a minority of these cells glow when Canseco was using?

Testosterone and other sex steroids can only bind to genes after they have first bound to the proper receptors. But different kinds of cells have different kinds of receptors. So testosterone acts as a transcription factor only in those cells with androgen receptors. These androgen receptors reside in the gelatinous material inside the cell called cytoplasm. It is the testosterone-androgen receptor complex that moves from the cytoplasm to the cell nucleus and actually binds to the gene, thereby activating it. So we can largely predict which cells would glow when Canseco was using steroids by the presence or absence of androgen receptors. Some of the notable cell populations, or tissues, with androgen receptors are in the skin, skeletal muscles (biceps, triceps, and so forth), testicles, and the prostate gland. There are also androgen receptors in various parts of the brain, including the hypothalamus (which controls libido, among other drives) and the limbic system (which regulates emotions including aggression).[1] So these are the cell populations where the androgen-sensitive genes will light up. The same genes elsewhere in the body will remain dim. This is the most basic way in which a gene's activity is controlled by the cellular environment.

Canseco's unnaturally high testosterone levels not only caused the shutdown of his testosterone production in the short term; in the long term, they caused a reduction in the androgen receptors in the androgen-sensitive cells. As such, ever higher doses were required to achieve the same effect on the muscles. And testicular shrinkage became an increasingly chronic condition, eventually permanent.

Even if we consider only Canseco's androgen-sensitive cells, there is great variation in the effects of testosterone on cells in, say, the triceps muscle, compared with cells in the testicles or brain. In the triceps, testosterone stimulates the growth and proliferation of muscle fibers; in the testes, testosterone stimulates sperm development. What causes these divergent effects? In part they are due to the fact that testosterone interacts with different receptors and hence activates different genes in the triceps cells and in the testes cells. But even in those cases where the same genes are activated, the effects can be very different, simply because the cells in which these genes are activated are different. The varied responses among androgen-sensitive cells are testimony for cellular control over gene actions and effects.

From the Cellular to the Social

The activity of a gene, the degree to which it glows, is called *gene expression*. The control of gene expression is called *gene regulation*. So far we have considered gene regulation solely at the cellular level, the level at which the gene is most directly regulated. But the cellular environment itself is influenced by both the surrounding cells with which it directly interacts and cells in distant parts of the body with which it communicates through the blood. So gene regulation is often initiated from remote sites in the body. Androgen-sensitive genes in muscle cells are regulated by androgens produced in the testes.

Some of the most fascinating forms of gene regulation are initiated outside of the body. Social interactions are a particularly important source of gene regulation. For example, in animals from fish to humans, the outcome of competitive interactions can influence testosterone levels, with all the consequent effects on gene activity.[2] So, too, can many other kinds of social interactions. When Canseco was informed of his trade while standing in the On Deck circle, his testosterone levels may have dipped. Though he didn't seem outwardly bothered when the ball missed his glove and bounced off his head and over the fence for a home run, inside his body it was another story. The activity of a number of his genes—not just those sensitive to androgens—was temporarily altered by this embarrassing event. And to the extent that psychiatric interventions can meliorate the psychological effects of traumatic events like these, these interventions will cause alterations in the regulation of genes in his brain. In fact, any alterations in Canseco's androgen levels resulting from athletic or other social interactions begin with alterations in gene expression in some of his brain cells. Let's consider now how social interactions could alter androgen levels through changes in the expression of genes in these brain cells.

Recall that Canseco was busted for gonadotropins (GT). Though his were ultimately derived from the placenta of pregnant women, most GT in women and all GT in men are produced in a tiny gland at the base of the brain called the pituitary. But the production and release of pituitary gonadotropin levels are controlled by a small group of neurons in the hypothalamus.[3] These neurons produce yet another hormone, called gonadotropin releasing hormone (GTRH).[4] GTRH stimulates the release of GT, which stimulates the production of testosterone, a system called the hypothalamic-pituitary-gonadal axis, which I will henceforth refer to as the "reproductive axis." If Canseco continues along his self-destructive path, he might eventually need to go further upstream along this

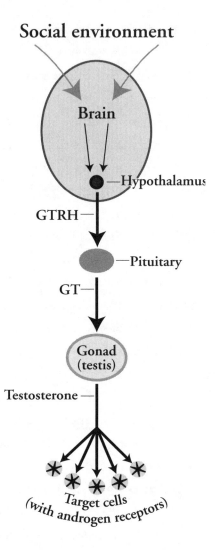

A schematic diagram of the hypothalamic-pituitary-gonadal (HPG) axis. Diagram by the author.

reproductive axis to GTRH itself, though that hormone will prove much harder to procure.

The only way that social interactions could impact Canseco's androgen levels is through his brain; that is, through effects on certain genes in certain brain cells. But the brain as we all know is a hugely complex organ. As such, we don't want to go on a fishing expedition for the target cells. Fortunately, we have a logical place to begin to unravel the mechanism for the social control of testosterone levels: those GTRH neurons. It seems likely that any social influences on the brain that impact testosterone levels will be channeled through these neurons. We can then work backward from the GTRH neurons to those neurons that provide inputs, either directly or indirectly, to the GTRH neurons. It is in those neurons that project to the GTRH neurons where we are most likely to find the initial changes in gene expression that are caused by the social environment.

Such studies would require experiments that no one would think of conducting on humans. Fortunately, we have animal models for these investigations. Surprisingly, one of the best animal models is a fish, specifically, the African cichlid *Astatotilapia burtoni* from Lake Tanganyika. Males of this species compete for territories, a prerequisite for attracting females. Only a minority of males in a population can maintain territories, however; the rest are relegated to nonbreeding status. The territorial males and the nonterritorial males look quite different. Territorial males have bold black markings on their faces and are generally more colorful than nonterritorial males. Internally there are even more dramatic differences. Territorial males have much larger testes and higher testosterone levels than nonterritorial males. The GTRH neurons are also much larger in the territorial males than in the nonterritorial males.[5]

The social status of these fish can be manipulated, however, such that the territorial males are transformed into nonterritorial males and vice versa, with all of the consequent external and internal changes.[6] The alteration in size of the GTRH neurons reflects in part a change in the activity of the gene that codes for GTRH.[7] A number of other genes are affected as well.[8] Of particular note, the genes for both the androgen receptor and the GTRH receptor become less active.[9] The net effect is a reduction in the rate of release of GT in the pituitaries of nonterritorial males, which causes a decrease in androgen production in the testes, with all of the effects we have previously detailed on androgen-sensitive genes, including those in the testes itself—hence Canseco-like shrinkage.

Lessons Learned from José's Sad Saga

There is a moral in José Canseco's misadventures with anabolic steroids, but it is only indirectly related to shrinkage. The moral pertains more directly to the remarkable sensitivity of genes to their cellular context. It is the cellular environment that determines how the genes for which testosterone is a transcription factor will respond. This sensitivity is not confined to testosterone-regulated genes; it is true of genes in general.

The cellular environment is itself influenced by other cells in the body, both local and remote. Moreover, the cellular environment is often influenced by events that occur outside of the body, including social interactions. So, many genes, including those responsive to testosterone, are ultimately socially regulated. The picture of gene action that emerges from Canseco's saga is one of dependence, not that of an executive instructing its biochemical minions. It is a story of genes as much directed as directors. But this is gene action over the short-term. Perhaps things will prove different if we consider gene actions over the longer term. Maybe then their sensitivity to context, from the cellular to the social, diminishes. Maybe over the long-term the traditional view of genes makes more sense. It is to these longer-term gene actions, in stress-related genes, that we now turn.

Chapter 4

The Well-Socialized Gene

THE EFFORT TO SECURE FUNDING FOR A VIETNAM WAR MEMO-
rial was inspired in part by a movie, *The Deer Hunter.*[1] The result,
despite the heated opposition of socially conservative philistines,
was Maya Lin's simple but powerful tapering black walls inscribed
with the names of the war's U.S. victims. *The Deer Hunter*, however, was less about those whose names appear on the Vietnam Veterans Memorial, than about those who survived the war but were
physically and/or psychologically wounded. In this chapter we will
focus on the nature of war's psychological wounds, so effectively
depicted in this movie. Some of these wounds are epigenetic in
nature. One of the reasons war and other forms of trauma have
such enduring psychological effects is that they induce epigenetic
alterations that cause long-term changes in gene regulation.

The Deer Hunter was released in 1978. It immediately resonated
with an American audience in the process of reconciling itself to
the horrible misadventure that was the war in Southeast Asia. The
events in the film occur in the late 1960s, when the war was at

its zenith and the country was riven by protests and counterprotests, which reflected a sociocultural, as much as political, divide in this country. The protesters were generally middle class and either in college or college educated. The counterprotesters were largely lower middle class blue-collar workers, who went to work full time right after high school. The protagonists in the movie come from the latter group. They are steelworkers in an unattractive small town south of Pittsburgh.

Michael (played by Robert De Niro) is clearly alpha, a man for whom leadership comes naturally; he embodies the kind of virtues that many Americans celebrate and deem distinctive to our culture: Decisive, action oriented, physical—the traits that helped make John Wayne a movie star and George W. Bush a president. But unlike John Wayne and George W. Bush, Michael has a reflective side. Steven (John Savage) functions as a sort of middle brother; he is loving and easygoing, about to marry a woman who is pregnant by another man. Nick (Christopher Walken) is the youngest, closest in age to the protesters. He is also distinctive in his introspection. He has an artistic sensibility that seems out of place in this group, in this town. Early in the movie, during the deer hunt, Nick says to Michael that he loves to hunt because he "loves the trees," the way each tree is different and unique. The relationship between Michael and Nick resembles that of an older brother and a younger brother but without any of the familial complications. Michael both understands and appreciates Nick's artistic sensibility. During the hunt, he tells Nick that "without you, Nicky, I hunt alone."

Before the opening scene, the three friends have decided to enlist in the army—Michael, for his own reasons, Steven and Nick because Michael did so. They are soon to be deployed, but before that there is the matter of Steven's wedding. It is a traditional Rusyn (an Eastern Slavic ethnic group) Orthodox wedding that reinforces for the viewer the fact that these men are only a generation or two removed from immigrant status. During the raucous reception, Nick asks his girlfriend, Linda (Meryl Streep), to marry him. She agrees. Michael, who is secretly attracted to Linda, gets righteously drunk later that night and runs naked through the streets of town. Nick eventually subdues him; he then makes Michael, drunk though he is, promise not to leave him "over there," if something happens. His plea has literal and metaphorical sobering effects on Michael.

Early the next morning, Nick and Michael along with three other friends—but not Steven, who is on his honeymoon—embark on their deer hunt. The other three friends are not interested in hunting so much as getting drunk; one even forgets his boots, which greatly angers Michael, a serious hunter. In fact, for Michael, hunting has a sacramental element and the deer is a somewhat totemic figure that must be treated with respect. Hence his obsession with killing the deer with "one bullet, one shot," which he proceeds to do.

The film jumps to a battle scene in a small Vietnamese village, during which Michael burns an enemy combatant with a flame thrower, then shoots the incinerated corpse numerous times with his M16. Reinforcements arrive, including Steven and Nick. Soon after the three friends are reunited, however, they are captured and held as POWs at a primitive facility on the edge of the Mekong River. For entertainment, their guards force the men to play Russian roulette, betting on the outcome. This is the second most crucial scene in the movie after the deer hunt.

Steven is the most outwardly, demonstratively terrified and traumatized by the prospect, so Michael focuses his attention on comforting him, hugging him, exhorting him to be strong. Steven is also chosen as the first of the three to "play." Meanwhile, Nick cowers in quiet terror, unconsoled. Eventually Michael convinces Steven to pull the trigger. Fortunately, he aims high, as the chamber was loaded. For his transgression, Steven is placed in an underwater wooden cage, the top of which he must grab to keep his head above water. During a hiatus, Michael convinces Nick that their best chance is to play each other with three (of six) chambers loaded, rather than one. If they both survive they will turn on the captors. But now, of course, the odds of surviving are much lower. Nick is reluctantly persuaded of the plan. When the gun is spun, it points to Nick, so he is the first to go. After much hesitation during which his captors scream threats, he pulls the trigger; the chamber is empty. Now it is Michael's turn. The tension is almost unbearable for both him and Nick. Michael steels himself and pulls the trigger. There is only a click. He immediately turns the gun on his surprised captors, and with the help of the rifle of the first captor shot, the two men manage to kill their remaining captors.

Michael, while plotting the escape, had advocated leaving the psychologically broken Steven behind, but Nick vehemently protests, so they bring him along. Once they escape the prison, they float down the river on a log until a rescue helicopter arrives. Nick is first to board, while Michael and Steven still dangle from the landing rails. Steven loses his grip and falls back into the river, Michael then lets go his own grip to rescue Steven. Michael manages to bring him to shore, then carries the now paralyzed Steven through the tropical forest until they reach a friendly convoy retreating from battle.

We next find Nick recuperating in a Saigon military hospital, showing signs of psychological damage. He can barely speak to his doctor. He does not know where his friends are. He may feel abandoned; he may feel survivor's guilt. In any case, he is isolated and alone. He wanders the red light district of Saigon at night, where he is introduced by an expat Frenchman to a gambling den, where they play Russian roulette. Michael is in the room and suddenly recognizes Nick, but his call to his friend, as he is being whisked away, is unheard.

In the next scene, Michael is returning home, believing that Steven and Nick are dead or missing. He finds the attention and support of his friends unwelcome. He retreats into himself. When he again goes out hunting, he tracks down a trophy buck but purposely misses high. From a rocky ledge he shouts, "OK?" as if to God, but hears only echoes in reply.

Unbeknownst to Michael, Steven is convalescing in a nearby Veterans Hospital; he is partially paralyzed and has lost both legs. When Michael learns of this, he visits Steven at the overcrowded VA hospital. The reunion is marred by the fact that Steven doesn't want to return home to his wife, family, and friends. Steven also informs Michael that he has been receiving large amounts of cash from someone in Saigon. Michael realizes that the cash is coming from Nick. He brings Steven home against his will, then departs for Saigon and arrives just before its fall in 1975. He eventually locates the French expat, who reluctantly takes Michael to Nick, his cash cow, in a seedy and crowded den. But Nick doesn't recognize Michael, nor remember much about his life in Pennsylvania. A desperate Michael enters a game of Russian roulette against Nick, talking all the while about Pennsylvania, to no avail. But his talk of past hunting trips finally resonates. Nick recognizes Michael, smiles, and says, "One shot." He then shoots himself in the head to Michael's—and our—horror.

One of the strengths of the movie is the varied ways in which Michael, Steven, and Nick respond to their traumatic experiences in Vietnam; they represent a microcosm of sorts of the reactions of Vietnam veterans, or the veterans of any armed conflict, as a whole. Michael, like many—perhaps the majority—of veterans, experienced temporary depressive symptoms (and perhaps lifelong nightmares) of the sort that you would expect of any sentient being, given his experience. His response closely resembles responses typical of someone grieving the loss of a loved one. Steven suffered a more severe and long-lasting depression, as evidenced by his desire for social isolation, even from his loved ones. Nick's psychological wounds were the most severe, true posttraumatic stress disorder (PTSD), though that term was not invented until a couple of years after the film was released.

But the psychic wounds of all three men have one thing in common: a problematic stress response that is at least temporarily pathological. There are two basic ways in which the stress response can go awry. First, it can be overly sensitive, too easily triggered, and hence chronically overactive. The result is various forms of anxiety disorders and depression, the problem experienced to different degrees by Michael and Steven. Second, it can react too robustly in response to a stressor, in effect blowing the circuits. This problem is more typical of PTSD as experienced by Nick.

The Stress Response

Obviously, the three friends did not have identical experiences in Vietnam—only Steven experienced the underwater prison, for example. But we will ignore these differences for the purposes of this discussion. Given this idealization, how do we explain their varying reactions to this trauma? Some people would emphasize their genetic differences; others would focus on differences in the way the three characters were raised. Most everyone, no matter what their biases, would make an ecumenical bow in acknowledging that it's not solely genes or environment but some combination of both, a cursory suture of the nature-nurture divide. Here we will explore a more intriguing possibility—that their early environments may have caused their genes to react differently to the same stress.

It is the pathologies in the stress response that we usually think of when we think of the stress response. But the stress response is utterly vital to our normal functioning as well, a fundamentally adaptive process that evolved to help retain a physiological equilibrium when confronted with the challenges of our dynamic environment. One indication of the importance of the stress response is that it involves virtually all of our physiological systems, from reproduction to the immune response.

The fastest form of the stress response is often referred to as *fight or flight*, during which the heart rate increases, the blood vessels dilate, the liver breaks down the energy store of glycogen to glucose, the primary energy source for our cells. All of these events prepare the body for quick decisive action. So too do other responses involving the skin (perspiration), the immune system (repair of wounds), and the brain (arousal and vigilance). Fight or flight is the initial form of the stress reaction in response to acute stressors, from drunk drivers to bears in our path to being shot at during combat, and, yes, to Russian roulette. When the stressor is more chronic—bullying, joblessness, trench warfare, and so on— the stress reaction involves many elements of the fight-or-flight response but also longer-term changes in the activated systems.

The stress response is initiated in the brain and involves two distinct but interconnected systems. Here we will focus on the so-called stress (or HPA) axis, the basic structure of which resembles the reproductive axis discussed in the previous chapter: A population of neurons in the hypothalamus produces a hormone, corti-

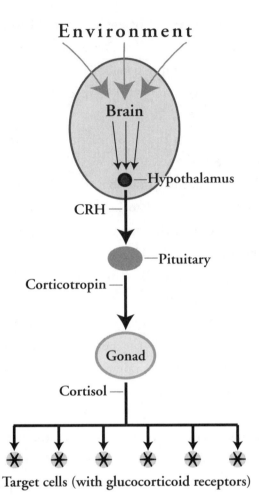

A schematic diagram of the hypothalamic-pituitary-adrenal
(HPA) axis. Diagram by the author.

cotropin releasing hormone (CRH), which stimulates cells in the
pituitary to release corticotropins (CT), which stimulate the adrenal gland to release glucocorticoid stress hormones including cortisol. Like the sex hormones testosterone and estradiol, cortisol is
a steroid hormone and influences gene expression by combining
with its nuclear receptor.[2] There are a number of glucocorticoids.

We will pretend, however, that cortisol is the only glucocorticoid, and that there is only one glucocorticoid receptor. Glucocorticoid receptors are far more abundant and distributed more widely than androgen receptors, and they activate a much greater variety of genes. For this reason, synthetic glucocorticoids, like cortisone, have even more side effects than does testosterone.

Stress-related pathologies occur when the stress axis is overtaxed, either by an overwhelming acute trauma like Russian roulette, or by chronic stress such as experienced by a soldier under constant threat. Whether the stress overload is acute or chronic, one of the more reliable indicators of stress is an elevation of CRH levels in the brain.[3] (Cortisol levels are also often elevated, but the relationship between a pathological stress response and cortisol level is more complicated.)

Of course, there is a lot of individual variation in how we respond to stress, as represented by Michael, Steven, and Nick. This individual variation has led to a search for genes that might explain this variation, such as genes for depression and anxiety. In a quite different approach, the emphasis is on events occurring early in life, beginning in the womb.

What Happens in the Womb Does Not Stay in the Womb

For decades, fetuses at risk for premature birth have been treated with a synthetic form of cortisol to promote lung development, because respiratory failure is one of the main hazards of premature delivery.[4] Recently, doctors and scientists have become concerned about the long-term effects of this treatment on the stress axis. These concerns are warranted. Fetuses that have received this treatment evidence lifelong hyperresponsiveness in the stress axis that results in a greater incidence of heart disease and diabetes, among other ailments, and a reduced life expectancy.[5] These treatments also predispose the recipients to stress-related brain/behavioral problems such as anxiety disorders, depression, substance abuse, and schizophrenia.[6]

These synthetic glucocorticoid treatments mimic the effects of maternal stress. When a mother-to-be is stressed, she produces more cortisol than she otherwise would. Some of this cortisol is transmitted to the fetus through the placenta. The elevated cortisol levels experienced by the fetus permanently adjust the settings of the stress axis of the fetus in a way that makes it more sensitive and hyperresponsive to subsequent stressful events. These permanent alterations in the stress response are often referred to as glucocorticoid, or HPA, programming.[7] Here I will simply call it "stress biasing."

A mother's stress could come from multiple sources. A bad marriage, social isolation, and poverty are just a few. Extreme stress levels, such as those thought to promote PTSD, can also result from diverse causes. War is a very effective promoter of PTSD, as so effectively portrayed in *The Deer Hunter*. Of course, the Vietnam War was not the first to produce victims of PTSD. Indeed, Herodotus, in 500 B.C., is alleged to provide the first account of this malady, in a veteran of the Greco-Persian wars who had witnessed the death of his close friend at the Battle of Marathon.[8] More recently, World War I resulted in many cases of "shell shock," and World War II, "battle fatigue," both less clinically euphemistic names for PTSD.

War is not a prerequisite for PTSD; any severe trauma will do. Natural disasters such as earthquakes, the 2004 tsunami in the Indian Ocean, and Hurricane Katrina are effective agents of PTSD.

Far from natural disasters, such as the 2001 destruction of the World Trade Center, have also caused PTSD.[9] The true toll of the Holocaust included not only the millions who were killed or died of starvation, but also many, many survivors who were permanently damaged in a way we now recognize as PTSD. In fact, this form of suffering caused by the Holocaust continues to ramify beyond its immediate victims, in a way that may be relevant to Nick's case.

Children of mothers who suffered PTSD as a result of the Holocaust are more prone to develop PTSD, even though they had no direct experience of the Holocaust.[10] Interestingly, though all children of Holocaust survivors are more prone to depression, second-generation PTSD is only elevated in those whose mothers suffered PTSD; there is no such correlation for children whose fathers experienced PTSD a result of the Holocaust. This fact suggests an important role for the fetal environment. The role of the fetal environment is especially evident in mothers who directly experienced the destruction of the World Trade towers. As you would expect, a number of them evidenced PTSD. Those who were pregnant at the time gave birth to babies with an elevated stress response and a hypersensitive stress axis.[11] They will be more susceptible to anxiety, depression, and even PTSD than those whose mothers did not experience PTSD. We would expect, then, that such traumas experienced through the womb could be a contributing factor to the susceptibility of veterans like Nick to PTSD.

PTSD is but an extreme case of a stress response gone bad, and perhaps the least well understood. Far more pervasive are stress-related pathologies such as those evidenced by Michael and especially Steven. It is to these less extreme pathologies, such as anxiety, fearfulness, and depression, that we will now turn. To get to the bottom of the mechanisms underlying these pathologies, however, we need nonhuman animal models for the necessary experiments.

For in utero effects of stress, the guinea pig is the animal of choice because, like humans, guinea pigs have long pregnancies and the young are born in a developmentally advanced state compared with mice and rats.

As in humans, when pregnant female guinea pigs are treated with synthetic glucocorticoids, the stress response of their offspring can be permanently altered.[12] Moreover, when pregnant guinea pigs are subjected to stress during the phase of rapid fetal brain growth, the male offspring also exhibit an elevated stress response. There are noteworthy alterations in the brain and pituitary that accompany these changes.[13] In the brain, the levels of the cortisol receptor are reduced, particularly in the hippocampus, which indirectly modulates the neurons in the hypothalamus that secrete CRH (see the figure on page 40). We will explore the effects of maternal stress on the cortisol receptors.

Mothering: Beyond the Womb

Much of the research on the "programming" of the stress response has been conducted on mice and rats. These rodents are born at an earlier developmental stage than guinea pigs and humans, before the "rheostat" for the stress axis is set. This makes them more amenable to certain kinds of manipulation, the results of which can be readily monitored.

Long ago it was observed that when baby rats are removed from their mother's nest for lengthy periods, they become stressed out for life. If, on the other hand, they are regularly removed for brief periods and handled carefully by humans, their stress response is actually reduced relative to unhandled siblings. In part, this is due to the mothers' response after these separations. It turns out that after a brief removal, the mother licks the returned pup with a vengeance, whereas after a long separation the pup is treated more like a stranger and is lick deprived. The tactile stimulation provided by the lick grooming has a dampening effect on the stress response that is lifelong. Subsequently, it was discovered that there is natural variation in the amount of licking a mother allots to her pups; some mothers are much better lickers than others. Pups mothered by good or generous lickers show a dampened stress response relative to pups mothered by poor or stingy lickers.[14] Moreover, if you take a baby mouse from a poor-licking mother and place it in a litter of a good licker, its stress response more closely resembles that of its foster mother than of its biological mother.[15] This is the background for the following experiments. Much of this research was conducted by Michael Meaney and his collaborators at McGill University.

Meaney found that adult offspring of good lickers have more glucocorticoid receptors (GR) in certain parts of the brain—especially the hippocampus—than the offspring of poor lickers.[16] This results in greater negative feedback sensitivity to cortisol and hence reduced levels of CRH. The reduced CRH levels result in a dampened stress-axis response to stressors relative to the offspring of poor lickers. Since these differences occur in adult offspring, they must result from long-term changes in gene regulation, above and beyond any short-term changes in gene regulation by cortisol of its target genes.

When the offspring of good lickers and poor lickers are cross-fostered as described above, the effects are reversed: the biological offspring of poor lickers raised by good lickers resemble the biological offspring of good lickers in all respects, including the number of glucocorticoid receptors in the hippocampus.[17] The reverse is also true. The cross-fostering experiments provide compelling evidence of a direct relationship between maternal care and the stress response, as manifest in GR levels, of adults.

What causes these long-term changes in the GR level? One obvious possibility is a long-term change in the cortisol receptor gene itself. To see how that might occur, we need to look upstream to factors that influence the expression of *GR*.* On the control panel of *GR* is a binding site for a transcription factor called NGFI-A (nerve growth factor inducible factor A). (For convenience we will reduce this unwieldy acronym for an even more unwieldy term to just NGF.)

When NGF binds to *GR*, it has an activating effect, increasing *GR* transcription. NGF levels are higher in pups mothered by good lickers than in pups of poor lickers.[18] There are no such differences in NGF expression in adult mice, however. So the early, transient difference in NGF expression must permanently alter the responsiveness of the cortisol receptor gene in the brain. This occurs by means of an epigenetic alteration of the *GR* gene.

Epigenetic Gene Regulation

There are a number of epigenetic gene regulation mechanisms. One of the most pervasive and well studied is called methylation, which occurs when a methyl group (three hydrogen atoms attached to a carbon atom, or CH_3) attaches to DNA.[19] The effect of methylation is to inhibit the expression of the gene to which it is attached. Unlike testosterone, cortisol, and the other transcription factors we have discussed, methylation is not transient; the methyl group tends

---

*Genes are often named for a protein for which they code. To avoid confusion, the convention is to italicize gene names (in this case *GR*), while GR refers to the protein.

to stay attached to the DNA, even after it replicates during cell division. The methylated DNA not only persists throughout the life of the cell but is transmitted to all of the cells that descended from the cell in which the original epigenetic change occurred. As such, those genes that are epigenetically turned off as a result of methylation tend to stay turned off in that cell lineage.

But at certain critical periods in early development, things are in flux methylation-wise. Some biochemical pathways promote methylation, and other biochemical pathways prevent methylation or even cause demethylation. In mice the quality of mothering, as manifest in licking, tilts things toward one biochemical pathway or another for the *GR* gene. Good mothering promotes the demethylation pathway, while bad mothering leads to methylation. When *GR* is methylated, the transcription factor NGF does not bind well; as a result, fewer GR proteins are produced in the hippocampus and the stress axis becomes hyperactive, predisposing the mouse to fearfulness and anxiety.

Given the nature of epigenetic regulation, the earlier methylation occurs during development, the more pronounced and pervasive the effects. But methylation and other epigenetic processes continue long after birth—indeed, throughout the lifespan. Epigenetic alterations, some of them occurring long after birth, are thought to account for many differences in identical twins with respect to their stress response. Twins, especially twins reared apart, often differ markedly in their stress response, as manifest in anxiety, depression, and PTSD.[20] Even twins reared together have increasingly divergent experiences as they age. To the extent that this divergence causes epigenetic differences, we would expect to find physiological and behavioral differences in the twins. If, for example, Steven had an identical brother, Stan, who stayed home to work at the steel mill, Steven would be more likely to have had an elevated stress response than Stan upon his return. Moreover, Steven would be more likely to have an elevated stress response ten years later.

Epigenetics and *The Deer Hunter*

The saga of Michael, Steven, and Nick, as dramatically depicted in *The Deer Hunter*, is a microcosm of the human reaction to extreme stress and some of the pathologies in the stress response that often result. All of these stress-related problems are caused by changes in gene regulation, which are long-term, sometimes lifelong. Such long-term changes in gene regulation are epigenetic. Here we explored one type of epigenetic process, called methylation, and one particular gene, which, when methylated, can result in a lifelong elevation of the stress response in mice. Scientists refer to results such as those from the mouse studies discussed here as a *proof of concept*. That does not mean that we can straightforwardly extrapolate from the mouse studies to the conditions of Michael, Steven, and Nick, or for that matter to any nonfictional human. That would be rash. For example, there are potentially a number of other genes, the disregulation of which could play an important role in a hyperactive stress response. And methylation, as we shall see, is but one type of epigenetic process. The mouse studies do suggest that research into epigenetic regulation will be a fruitful avenue on our way to understanding stress-related pathologies and ultimately to treating them.

Chapter 5

Kentucky Fried Chicken in Bangkok

ON A VISIT TO THAILAND IN 2001, I SPENT THE FIRST DAY,
while suffering the miasma of jet lag, at the National Museum
complex in Bangkok. Though I am a museum lover, I predictably
found it hard to pay much attention to the displays and especially
to the captions. After a half-conscious hour or so, my stomach
churning, I found a welcome and therapeutic diversion in people
watching. Particularly engrossing were the multitude of local school
children visiting the museum on field trips. Most were dressed in
particularly drab school uniforms—khaki shirts and short pants,
dark brown socks worn high, and lighter brown shoes. The color
of their attire seemed somewhat incongruous in this tropical city,
though the shorts were quite practical.

But more noteworthy was their behavior. No shouting or
screaming or running wildly up and down the galleries. They were
quite attentive to their instructors and could be efficiently mus-
tered when it was time to move as a group. Even upon exiting the

museum to the crowded and chaotic streets, they retained their single-file formation and poised demeanor.

My own son was ten at the time, so I could not help but notice the differences regarding his own school outings, which were always attended by multiple chaperones, who, as I can attest from my own experience, were invariably exhausted by the end of the day from their only partially successful attempts at maintaining order.

On subsequent trips to Thailand, I would spend a couple of days in Bangkok at the beginning and end of the trip continuing my informal study of the school kids while visiting various temples and palaces. My focus gradually shifted, though, from their behavior to their physical attributes. Most obvious was their height, which was much closer to American standards than their behavior. As is true throughout much of Asia, each generation since the last world war has experienced a size increase. Many twelve- and thirteen-year-old Thai kids are already taller than their grandparents and will soon exceed the height of their parents. There are obvious reasons for this, most notably enhanced nutrition, especially protein.

Weight-wise, however, the Thai youth were still lagging far behind their American counterparts. On my first visit, one boy had stood out. I'll call him Paradorn. He was the only Thai kid in the museum who was clearly overweight. On subsequent visits I noticed more and more overweight boys (but not girls). By 2005 there seemed to be about one per class of about twenty-five kids. This is about the ratio I remember from my own grade school days in the 1960s: about one clearly overweight kid in each of my class pictures. That was much less than the obesity rate in my son's classes (for both boys and girls).

Weight-wise, the Thai kids more closely resembled American kids of the 1960s than those of my son's age. It is only recently that any preadolescent Thai kids have become overweight, and obesity seems to be restricted to the more urban areas; you will still have to search long and hard to find a fat Thai child in the Issan region, for example.

There is an obvious cause for this urban-rural weight difference in Thailand. Urban Thais are markedly more affluent that rural Thais; this discrepancy is the main cause of the recent political unrest. With affluence comes increased calories and often less physical activity—the standard formula for weight gain. But the source of the increased calories also matters, and here, too, urban Thais differ from their rural counterparts.

The diet of my friend Aniwat, though he lives in Bangkok, is typical of the rural village where he grew up, near Kaeng Krachan National Park in Petchburi Province: mostly a diverse range of fruits and vegetables, some of which he acquires at open-air markets, some he forages himself. On every walk in the forest, Wat would gather morning glory and other greens, as well as the fruits of countless species of eggplant, none of which looked much like the large, elongate, purple-skinned stuff that I once grew.

He eats much less animal-derived protein than a typical American, and virtually no beef. Chicken and fish are the primary animal proteins consumed by rural Thais, augmented by a diversity of creatures most Americans would not consider edible, notably a variety of insects, from grubs to cicadas and cockroaches. (For Wat, a green papaya salad without a cockroach is not a proper green papaya salad.) Wat consumes virtually no dairy products or processed foods, though he did acquire a taste for peanut butter during an extended visit to the United States. His desserts largely consist of processed fruit confections that are not particularly sweet. At age sixty plus, he retains the physique of the Muay Thai boxer he once was.

The diet of urban Thais, especially those recently arrived from the hinterlands, encompasses many of these elements—including the insects—but also a much greater amount of prepared foods, both Thai and increasingly European and American. The latter comes primarily in the form of fast-food chains—McDonald's, KFC, and such—which provide the most efficient path to obesity known. There are some interesting peculiarities in Thai fast-food preferences. Thailand is one of the few places I have visited where there seemed to be more Kentucky Fried Chicken franchises than McDonald's. In part this can be explained by the Thai preference for chicken over beef. But according to Wat, there is also the "fried factor." Much traditional Thai food is fried, albeit in a wok and with much less oil. That predisposition toward fried foods would also explain the recent success of doughnut purveyors in urban Thailand.

Originally, I attributed the accelerating weight gain in affluent urban Thais to their Americanized eating habits, a trend that began when the United States used Thai air bases as staging areas for attacks on Vietnam, Laos, and Cambodia from the mid-1960s through the early 1970s. Thailand is also the place where physically and psychically exhausted soldiers went for R&R. Naturally, they were looking for American food and they got it. An increasing number of Thais acquired a taste for American food in some of its most unwholesome incarnations—and they gained weight. When you eat like an American, you get fat like an American. That all seemed pretty obvious. Subsequently though, I have come to understand that the obesity of neither Americans nor Thais, like Paradorn, is quite that simple.

Though it is certainly true that McDonald's, KFC, and the like are major contributors to obesity in the United States, and more recently in places like Thailand, these fast-food outlets should be viewed more as a precipitating cause, the cause that tips a predisposed body over the edge. What is it that makes a body pre-disposed? The conventional answer is genes. Some individuals or ethnic groups are predisposed toward obesity because of their bio-logical inheritance. In this chapter we will explore a different sort of predisposition: epigenetic. These epigenetic predispositions gen-erally develop in the womb or in infancy.

Thrifty Genes?

Obesity per se is not a public health problem; it's the bad things that obesity does to your physiology, notably the so-called meta-bolic syndrome, which is basically a problem in the way your body processes food and can lead to cardiovascular disease and diabe-tes. It was in an attempt to explain diabetes in particular that the "thrifty genes" hypothesis was proposed.[1] James Neel noted in the early 1960s that when non-European populations were exposed to a Western diet, the incidence of diabetes (and obesity) skyrocketed to levels that were much higher even than those in the United States. He proposed, by way of explanation, that these populations had evolved in an environment of periodic famine in which individuals whose bodies were particularly efficient at turning calories into fat stores were at a selective advantage. These individuals could thrive during lean times because of their "thrifty genes," but they become fat and diabetic in an environment where food is plentiful.

The thrifty-genes hypothesis was criticized on a number of grounds. Most damningly, there is no evidence that humans expe-rienced periodic famine until the agricultural revolution 9,000 years ago.[2] Neel himself soon abandoned the idea, but it survives to this day in mutated forms.[3] The thrifty-genes hypothesis reflects a genocentric view of the obesity epidemic, a view that is also manifest in the search for obesity genes.[4] There is no lack of candidate obesity genes, but none of the candidate genes, alone or combined, go very far toward explaining who gets fat and why.[5]

While the gene hunters were busy at their labors, other researchers approached the obesity issue from a different angle, based on the fact that Americans and many Europeans were getting fatter at an accelerating rate. It was hard to pin this obesity epidemic on obesity genes. Instead, it is widely recognized that the American diet was too much of a muchness, and that the physical activity of the average American was insufficient to burn all of those calories. This all seems obvious enough, conventional wisdom even.

The Western lifestyle certainly provided a better explanation for the fact that many non-Western peoples, like Inuits, Pacific Islanders, and urban Thais like Paradorn, experienced dramatic weight gain when exposed to McDonald's and Kentucky Fried Chicken. They were just newly experiencing essentially what native-born Americans had been experiencing for decades. Whatever genetic differences may have existed among these groups, the signal was quite faint compared with that of diet and physical activity.

But what about the individual variation within these groups? Paradorn is still the exception among urban Thais. Moreover, not all Americans of European descent are overweight; in fact, there is tremendous variation in weight within this group. While most of this variation can be attributed to lifestyle factors, individuals with essentially the same eating and exercise habits can differ substantially in their weight. It is this variation that currently motivates most of those searching for obesity genes. The gene hunters' reasoning goes something like this: Even if we account for diet, exercise, and so on, there still remains the fact that people vary in weight and from an early age; ergo, they are born with different predispositions to obesity; ergo, they are genetically different in their propensity to gain weight. On this view, Paradorn got dealt a bad genetic hand.

But this seemingly obvious chain of logical inferences is only as strong as its weakest link. And its weakest link is the one that connects "ergo, we are born with different predispositions to obesity" to "ergo, we have genetic differences." That is simply a non sequitur, one you often find in discussions of gay genes, intelligence genes, and genes for a host of other human attributes. We cannot infer that there is an obesity gene or a gay gene or any other sort of gene from the fact that we are born with a predisposition toward obesity, homosexuality, or whatever. As we saw in the previous chapter, many of us are born with a predisposition toward a hyperactive stress response that is not there at conception. This predisposition develops in the womb as a result of epigenetic processes. The same may be true of obesity and its associated disorders in people like Paradorn.

From Thrifty Genes to Thrifty Phenotypes

Of course, our genes affect our weight. At issue is whether, in coming to grips with the obesity epidemic, our research dollars are best spent searching for particular mutant alleles at particular genetic loci. Obesity is not a simple trait like Huntington's disease, which can be traced to a mutation in a single gene. Rather, obesity is influenced by any gene that affects the way we process food, of which there could be hundreds, each having a fairly small effect.[6] So the gene hunters' task is to identify variant or mutant alleles among this large number of genes that may contribute slightly to obesity. This is a huge task under any circumstances, the payoff of which is uncertain at best.

Meanwhile, a quite different research program has proven productive. The objective in this alternative approach is to identify events in the womb that might result in obesity. It has long been common knowledge that events in the womb affect the health of infants, hence the emphasis on prenatal care. But this research greatly extended the range of conditions that we know are more or less directly affected by the in utero environment and their duration, not least obesity and other elements of the metabolic syndrome.

As was evident from the Dutch famine, one indicator of the quality of the womb environment is birth weight. Low birth weight generally indicates poor conditions in the womb. Not surprisingly, neonates born with a below-average weight are subject to a host of health disorders during early infancy. What is surprising is that these individuals continue to be less healthy throughout their lives and have a lower life expectancy as a result.[7] As we saw in Chapter 1, one of the adverse consequences of low birth weight is obesity in adulthood. Why would a small neonate become an overweight adult? The current consensus is that this association results from a process called fetal programming,[8] much of which occurs in the womb.

James Barker proposed that when the fetus receives insufficient nutrition through the placenta, it becomes programmed in the womb for a thrifty phenotype.[9] As was proposed for the thrifty-genes hypothesis, those with a thrifty phenotype have a more efficient metabolism than babies born at a normal birth weight. But the thrifty phenotype can result from diverse genetic backgrounds and without the aid of specific obesity genes. It is, rather, simply a function of the intrauterine environment.

The thrifty phenotype works out well in traditional non-Western cultures where the postnatal environment is often one of scarcity. In those cases, the prenatal environment predicts the postnatal environment in an adaptive way. Problems arise, however, if the postnatal environment is enriched food-wise relative to the

prenatal environment. When this mismatch occurs, thrifty phenotypes result in obesity and its consequences. The Barker hypothesis
nicely accounts for the correlation between low birth weight and
adult obesity and has been bolstered by much, though certainly not
all, subsequent research.[10]

But what is the nature of this so-called programming?

Barker himself has little interest in the mechanism through
which this effect of the uterine environment occurs. Others,
though, have pursued mechanistic investigations of thrifty phenotypes, and, as is usually the case, the initial studies have been
conducted on nonhuman mammals—especially mice, rats, and
sheep. For studies of this sort, biologists often look for changes in
the expression of particular genes, as evidenced either by the abundance of their associated proteins (translation products) or messenger RNA (transcription products). In this case, they are looking
for long-term differences in gene expression between those individuals born underweight and those born at a normal birth weight.
And indeed, scientists have found a number of differences in gene
expression associated with birth weight.[11]

Many of these gene expression differences are tissue specific. For
example, one gene may be more (or less) active in the liver of a small
neonate, while another gene may be more (or less) active in the
adipose (fat cells) tissue. Other genes, notably the glucocorticoid
receptor gene (*GR*; see previous chapter), show different expression
patterns in many tissues, including several parts of the brain, the
liver, the adrenal gland, the heart, and the kidney.[12] These different
gene expression patterns often persist into adulthood and old age.

The products of many of the genes that vary in expression
according to the uterine environment are transcription factors, each
of which influences the expression of many other genes. The net
result is a host of long-term differences in gene expression in many

tissues that relate to the intrauterine environment. The challenge is to sort out the cause-and-effect relationships in these gene expression patterns and to causally connect them to events in the womb. Because these gene expression differences are long-term, researchers have begun to search for epigenetic signals.

Methylation Patterns Vary with Diet

The expression of one family of genes, in particular, seems to be directly connected to nutrient availability in the womb: the genes that code for DNA methyltransferases (Dnmt).[13] Dnmt promotes and maintains methylation in genes subject to epigenetic regulation. Hence, when Dnmt levels are high, these genes tend to be turned off, or silenced. Conversely, low Dnmt levels and the consequent reduced methylation result in increased expression of those genes.

In rats fed a protein-restricted diet during pregnancy, *Dnmt* expression is low.[14] Low Dnmt levels mean that some genes that should be methylated are not. Because these genes are not methylated, we would expect them to be more active than they would normally be, producing more of whatever it is they produce. One of the genes methylated by Dnmt is *GR*.[15] This gene is another example of tissue-specific (and hence context-sensitive) gene regulation. Recall that in the hippocampus, NGF binds to *GR*. In the liver, on the other hand, it is Dnmt that binds to *GR*, thereby deactivating it. In rats fed a low-protein diet, Dnmt levels drop, resulting in lower methylation of *GR* and hence increased *GR* expression. Just as lower-than-normal levels of *GR* expression in the hippocampus cause problems such as a hypersensitive stress response, higher-than-normal levels of *GR* expression in the liver also cause problems. The higher-than-normal levels of GR in the liver and other tissues cause these tissues to be too sensitive to stress hormones. The long-term effect is an increased risk of diabetes, obesity, and other elements of the metabolic syndrome.[16]

The link between *GR* expression and the metabolic syndrome has led some to speculate that ultimately, the low nutrient levels in the womb are just another sort of stressor, the effects of which are mediated by the stress response.[17] If so, other forms of stress in utero that result in high cortisol levels should mimic the effects of bad nutrition. In fact, there is evidence that social stress experienced by the mother increases the chances that her offspring will develop the metabolic syndrome. A fetus subject to both kinds of stress, as occurred during the Dutch famine, would be especially vulnerable to the metabolic syndrome.

Maternal stress adds another potential dimension to the obesity epidemic. Some researchers have proposed that increased obesity levels are in part attributable to the highly stressful Western way of life, especially in urban settings.[18] This stress is transmitted to the fetus through the placenta, resulting in obesity, diabetes, and so on. Thus it is possible that Paradorn is overweight because of the stress his mother was under while pregnant. That stress could have a number of sources, of course—poverty, for example. Or it may have been her social environment. Perhaps she moved away from her family in the rural Issan region for a better life in bustling Bangkok. That not only would amount to severe culture shock; it would have left her isolated, without the social support of traditional rural Thai families. Whatever the source of her stress, it may have affected both Paradorn's weight and stress response.

It is worth noting as well that too much of a good thing in the womb can also be a stressor. A fetus that gets too many calories also has an elevated stress response and is more prone to obesity.[19] Perhaps it is for this reason that overweight neonates, as well as their underweight counterparts, tend to become overweight adults. For Paradorn, that would mean a quite different explanation *vis-à-vis* his mother as it relates to his plight. Having moved to the city, she abandoned her traditional diet for McDonald's and KFC, and her cravings for these foods only increased during pregnancy. The effect of these excessive calories on Paradorn—whether directly through its effect on his metabolic rheostat, or indirectly through his stress response—was to predispose him toward obesity. This purely speculative scenario is intended solely to provide some sense of the potential diversity of environmentally induced epigenetic changes relevant to obesity.

From DNA to Histones

So far, we have only considered one avenue by which methylation exerts its epigenetic effect on gene activity: through the binding of methyl groups to or near a particular gene. But many of the effects of methylation on gene expression are more indirect; these indirect effects come by way of a class of proteins called histones.[20] There is evidence that diet-induced histone modifications in the fetus are a factor in the metabolic syndrome.[21]

When I first learned about DNA in high school biology, my mental picture of things at the molecular level had the naked double helices sort of floating around the nucleus, always at the ready for protein synthesis. It was only later, and with some mental effort, that I came to understand that DNA is far from naked but, rather, intimately entangled with proteins. It is this DNA-protein complex that constitutes chromosomes. DNA and protein are so entangled that, as mentioned in Chapter 2, after the discovery of chromosomes, it was quite unclear whether it was the DNA or the proteins that were the genetic material. Naturally, once DNA was proven to be the genetic material, the protein component of chromosomes was largely ignored.

The proteins in chromosomes were thought to function primarily in efficiently packaging inactive DNA in a condensed state, one that occupied far less space than that of the expanded, active form of DNA—sort of like archiving a computer file. It is only recently, primarily as a result of epigenetic research, that a quite different view of these chromosomal proteins has emerged. On this new view, histones are much more dynamic than previously supposed and they play an important role in regulating gene expression.

In general, histones are less tightly bound to the DNA where the genes are actively engaged in protein synthesis, and more tightly bound to the DNA where genes are inactive. The degree to which a histone is bound to the DNA is a function of epigenetic processes. These histone-related epigenetic processes involve various types of biochemical alterations to the histone, one of which is methylation.[22] As in DNA methylation, histone methylation usually (but not always) blocks gene expression. And as with DNA methylation, histone methylation is passed, intact, from a cell to its descendents.

Rats that experience low-protein diets during development evidence histone modifications near the *GR* gene that cause it to be expressed at higher-than-normal levels.[23] Whether these histone-based alterations precede or follow alterations in the expression of this gene caused by DNA methylation is unclear. DNA methylation and histone methylation are often coordinated. For the purpose of fine-tuned therapeutic treatments, it is important to know, in a precise way, just how these two forms of methylation are coordinated. Currently, it is known that folic acid and other key nutrients (for example, zinc, vitamin B12, and choline) can remedy somewhat the effects of poor nutrition in the womb through their epigenetic effects.[24]

Folic acid was first used as a prophylactic against spina bifida and other neural tube defects. It has proven quite effective in that regard when consumed by a mother-to-be in the first trimester. This effect of folic acid occurs via epigenetic modifications of certain key genes in neural development. Subsequently, it was discovered that folic acid has other epigenetic effects during development, some of which can meliorate the metabolic syndrome.[25] These epigenetic effects of folic acid extend well beyond birth, perhaps to adulthood. For this reason, most food manufacturers fortify all grain products from cereal to flour with extra folic acid (which is usually obtained from fruits and vegetables). This is perhaps the first application of nutritional epigenetics.

But there are reasons for caution with regard to this uncontrolled experiment. Given its potential epigenetic potency, too much folic acid could be a bad thing. Some suspect a link between high levels of folic acid and autism, for example, based on epigenetic considerations.[26] The putative increase in autism roughly corresponds to the time that folic acid became widely added to our foodstuffs and consumed in high doses by pregnant women. Moreover, epigenetic differences have been identified in some diagnosed with autism.[27]

At this point, the folic acid–autism link is almost purely conjecture. But there can be no doubt that nutritional epigenetics has a bright future, both as prophylaxis and therapy. As prophylaxis the big payoff will come when scientists can influence the "fetal programming" of obesity, diabetes, and other conditions, in people like Paradorn, through nutritional silver bullets, carefully timed and targeted. Therapeutically, the payoff will come from diets formulated specifically for those either susceptible to or experiencing these ailments—from childhood on. Both the prophylactic and therapeutic potential of nutritional epigenetics extends to many other ailments as well, such as cancer, which I will discuss later in the book.

What Predisposed Paradorn?

We have explored several possible explanations for Paradorn's weight; these can be divided into two broad categories: genetic and epigenetic. They are not mutually exclusive. It is possible that Paradorn represents a rare combination of thrifty and obesity genes, which caused problems given his food-enriched environment. Alternatively, Paradorn's predisposition may have originated in the womb or his early postnatal environment. If so, his predisposition would be largely epigenetic. I have discussed one possible epigenetic mechanism involving Dnmt and the *GR* gene, as they relate to nutritional factors and stress. The epigenetic and genetic explanations both involve genes but in fundamentally different ways. Genetic explanations for Paradorn's predisposition require sequence variation, that is, variation in the alleles at a particular genetic locus. Such variation is immune to environmental influence except via mutation. Epigenetic explanations for Paradorn's plight, on the other hand, involve variation in chemical attachments, either to key genes or to adjacent histones, which can be quite sensitive to the external environment.

One reason genetic explanations for obesity continue to garner so much attention is the observation that obesity runs in families. Epigenetic processes begin and end in a single lifetime. Or so it was thought. Recently, it has become apparent that epigenetic processes, including those involved in obesity, can be transgenerational as well. That is the subject of the next chapter.

Chapter 6

Twigs, Trees, and Fruits

I VISITED THE TORONTO ZOO FOR THE FIRST TIME ON A BRISKLY beautiful day in October, 2008. One of my first stops was the gorilla exhibit, where I remained entranced for over an hour. The social dynamics were fascinating, and there was a docent present, who augmented what we were seeing with much useful information about the relationships among the individuals, their life stories and personality quirks. There were several adult females, a couple of juveniles, two youngsters—one born fairly recently—and a silverback male, much like a gorilla group in the wild. Physically, the silverback, Charles, was the most riveting figure, with his huge head and neck and an upper body that would put any steroid-enhanced body builder to shame. Charles was nearly thirty-five years old at the time, past his prime for a gorilla in the wild but still good to go in the cage. He had sired numerous progeny, including the youngest juvenile, during his many years at the zoo.

Charles's personality, though, was much less compelling than his physique. Older silverbacks in general tend to be dour and not much prone to initiate interactions with the rest of the troop. But Charles is extreme in this respect, even for a silverback. For example, he has a strong-seeming aversion to his own young offspring, which is not normal. The females, both mothers and aunts, were acutely aware of this and tried to steer the ebullient youngsters away from papa, lest, presumably, they get a swat. This was quite entertaining to watch, as the youngest in particular was not easily deterred from interacting with the imposing fellow. The mother was often forced to physically divert her youngster, warily watching papa all the while.

Charles could perhaps be excused for his social infelicities, given his own traumatic early life. He was found in the wild beside the body of his dead mother, who had been shot by poachers. Soon thereafter he was sent to the Toronto Zoo, where he was raised by humans. He never had a chance at normal gorilla socialization. In retrospect, it is impressive that Charles was such a successful sire. Many human-raised male gorillas are sexually incompetent.[1] Some who do engage in sexual activity can't even do it right. In general, human-raised male gorillas don't make good sires. There are a lot of data on this because a considerable percentage of captive gorillas are rejected or neglected by their mothers and must be hand raised as a last resort, for both ethical and conservation purposes. (All three subspecies of gorilla are threatened with extinction.)

For hand-raised females, the consequences are more far-reaching. Social interactions are much more important for adult females than males; it is their interactions that provide the social glue for a group. Human-raised females often have all sorts of issues as a result of improper socialization. And one of the main issues is mothering, which is why there are so many neglected captive gorillas that need to be human nurtured. This creates a vicious cycle: human-nurtured gorillas make inadequate mothers, which results in more human-nurtured gorillas and hence more inadequate mothers.

This problem is exacerbated by the fact that for conservation purposes, the females are often moved around from zoo to zoo for pairings designed to minimize inbreeding. This makes captive female groups more unstable and hence more stressful for the constituents. It takes a group to raise a gorilla and stable groups are always better. The effect of confinement itself must also have some effect, if only because any disputes are less easily meliorated by, for example, avoidance. In any case, there is a huge mothering issue among captive gorillas.

Clearly, mothering is not an instinct in gorillas; it is a learned skill. But the motherless mothers problem also points to the fact that to be a good gorilla mother, she must be in the proper emotional state, and not having had a gorilla mother herself makes it less likely that she will be in that state when confronted with her own baby.

For the purposes of this chapter, I will focus on the fact that gorilla mothering also represents a form of inheritance, a social inheritance.[2] For gorillas, proper mothering requires a well-functioning social structure, which zoos have been unable to adequately replicate. The effects of these deviations from the social norm influence neural development and other physiological processes. These physiological changes are often mediated by epigenetic processes. In this chapter, we will explore epigenetically mediated social inheritance in gorillas and in other social animals from rodents to humans.

From Motherless Monkeys to Poorly Mothered Rats

In the 1950s, Harry Harlow conducted some pioneering experiments at the University of Wisconsin concerning the emotional attachment of an infant to its mother.[3] The experiments were conducted on rhesus monkeys, a favorite primate model for many scientific investigations. I remember watching a film on the experiments as an undergraduate and feeling both enthralled and repulsed, for certainly by today's standards, Harlow's experiments seemed to cross an ethical line in his treatment of his subjects. Harlow himself did not come across as a particularly sympathetic figure but as an almost Dr. Strangelove caricature of an experimental psychologist. But the claims of his most outspoken detractors—that his sometimes sadistic seeming experiments yielded nothing of scientific value—are complete nonsense.

Harlow's initial experiments were designed to address a fundamental question about the mother-infant bond from the infant's perspective: is it sustained by the sustenance the infant receives (initially milk), or by some other, less obviously life sustaining qualities that the mother has to offer? The answer may seem obvious now but it wasn't at the time. According to the then prevailing view of the behaviorist school, infants clung to their mothers only because of the food rewards she offered. Harlow doubted the behaviorist line and decided to test it. He constructed wire-monkey mothers, some equipped with nipples through which milk could flow, others lacking nipples but clothed in terry cloth. Neonates that had been recently removed from their biological mothers could choose which artificial mother to cling to. All the neonates quickly learned to get their milk from the bare-wire surrogate but spent the rest of the time, including sleep, clinging to the terry-cloth surrogate. The tactile stimulation provided by the terry-cloth surrogate—though a poor approximation of a mother's fur—was more compelling to these infants than the milk provided by the bare-wire version.

Obviously, an inanimate wire surrogate, even covered in terry cloth, is no substitute for an actively grooming and solicitous mother. So the infants "nurtured" by the surrogates exhibited high levels of stress and deep psychosocial problems. Nor could they ever be properly resocialized with other monkeys, no matter what methods were tried.[4] In the 1960s, one of Harlow's students, Steven Suomi, investigated the behavior of surrogate-raised females when they became mothers themselves—in other words, the motherless mothers. He found that they were at best neglectful and at worse abusive toward their own infants.[5] There are obvious parallels here with the captive gorilla situation. Moreover, these studies provided some important insights regarding the human condition, particularly the fact that child abuse and neglect tends to run in families over several generations. Neglect begets neglect. Abuse begets abuse.

But *how* does neglect beget neglect? What happens in the brain of a neglected infant to make it a neglectful parent? For an approach to answering that question, we turn again to the rat studies by Michael Meaney and his associates. Recall that mother rats vary in the degree to which they tactilely stimulate their offspring through licking, and that rats not receiving adequate licking tend to become stressed-out adults as a result of epigenetic changes in the NGF gene. What happens when these stressed-out adults become mothers? Pretty much the same thing, it turns out, as occurs in captive gorillas and Harlow's rhesus monkeys. Neglected (lick-deprived) female rats become neglectful mothers.[6]

Through the rat model, we can delve a bit deeper into the mechanism for this social inheritance. One possible explanation is that stressed-out mothers are neglectful mothers: A lick-deprived mouse becomes a stressed-out mother as a result of epigenetic alterations to the NGF gene; because of her stress, she is a neglectful parent; as a result, her female offspring experience the same epigenetic alterations to their NGF genes and become stressed-out mothers too—

and so the cycle continues. That seems to be part of the story, but not the whole story.

Mother rats, like all mammal mothers, including humans, undergo a suite of hormonal changes just prior to and after giving birth. Levels of oxytocin rise, as do levels of estrogen and estrogen receptors. Estrogen receptor levels seem to play a particularly important role in maternal behavior. Levels of this receptor are reduced in the female offspring of poor lickers relative to those of good lickers.[7] One consequence of the low estrogen receptor levels is that when the female pup becomes a mother, she doesn't respond normally to the elevated estrogen levels she experiences when she gives birth. One consequence of this dampened response is a reduction in the binding of oxytocin in the hypothalamus, a region of the brain that is crucial for maternal behavior. Oxytocin, especially in its actions in the hypothalamus, promotes social or affiliative behavior.[8] (Some researchers, reductively, refer to it as the "love hormone.") This effect of the estrogen receptor on oxytocin occurs because the estrogen–estrogen receptor complex stimulates the expression of the oxytocin receptor gene in the hypothalamus by binding directly to the oxytocin receptor gene's control panel.

The reduced expression of the estrogen receptor gene in female pups tends to persist into adulthood, hence making it more likely that she will be a less devoted licker when she becomes a mother, perpetuating for another generation the effects of inadequate maternal care. As you might expect by now, the long-term effects of lick deprivation on the expression of the estrogen receptor gene are epigenetic in nature. Pups born to good lickers but raised by poor lickers have lower levels of the estrogen receptor in the hypothalamus than their siblings that remained with the biological mother.[9] The reverse is also true: pups born to poor lickers but raised by good lickers have elevated levels of the estrogen receptor in the hypo-thalamus. In both cases, the alterations in estrogen receptor levels in the hypothalamus persist as a result of methylation of the con-trol panel for the estrogen receptor gene. Pups raised by poor lick-ers evidence more methylation of this control panel than do pups raised by good lickers. (Recall that high methylation levels gener-ally result in lower expression levels for a given gene.)

Social Inheritance

Female offspring of poor lickers experience an epigenetic double whammy that affects their own mothering. First, as we saw in Chapter 4, their stress response is elevated as a result of epigenetic alterations in the NGF gene in the hippocampus. Since the pres-ence of neonates is stressful in and of itself, they are distractible and unsolicitous in the presence of their own young. Second, as a result of epigenetic alterations of the estrogen receptor gene in the hypo-thalamus, they are less likely to lick their own offspring even when they are relatively unstressed.

Therefore, poor mothering tends to perpetuate itself in a vicious cycle; conversely, high-quality mothering tends to perpetuate itself in a virtuous cycle, from one generation to the next. This is a form of social inheritance mediated by epigenetic processes. Though most research on maternally based social inheritance has been conducted on rodents, there is substantial evidence for simi-lar processes in primates, including humans. In rhesus monkeys, the species studied by Harlow, maternal rejection and abuse during the first three months causes a multitude of pathologies in brain and behavior, including the stress response.[10] Similar effects have been observed in other primates.[11] In rhesus monkeys, under conditions much less severe than those of Harlow's experiments, maternal behavior, and hence its effects on offspring, tends to "run in families."[12]

Humans have an especially protracted infancy and childhood. Children subject to poor parenting, including both psychological and physical abuse, suffer reduced mental health.[13] As in rats and monkeys, these effects are associated with an altered stress response.[14] Moreover, also as in rats and monkeys, poorly parented children tend to grow up to be poor parents.[15] Recall from Chapter 4 that lick-deprived rats have an elevated stress response as a result of epigenetic changes to the NGF gene in the hippocampus. There is now evidence for a similar effect in humans abused during childhood.[16]

But parenting that falls far short of the abuse threshold can also have lifelong effects on behavior, much of it mediated by the stress response. The best documented of these effects in humans concern the quality of maternal care as measured by an index called the Parent Bonding Instrument, or PBI. Low scores for maternal care are often, somewhat paradoxically, associated with high levels of maternal control. The combination is referred to as *affectionless control*, which is a risk factor for depression, anxiety, antisocial personality disorder, obsessive-compulsive disorder, drug abuse, and a reactive stress response.[17] In contrast, high levels of maternal care, as measured by the PBI, are associated with high self-esteem, low anxiety, and a dampened stress response.[18]

"Maternal style," is the term sometimes used to describe the suite of behavioral responses of a mother to her offspring.[19] The term encompasses not only abuse and neglect but also what would be considered the normal range of maternal behavior, from affectionless control to an affectionate hands-off approach, and all manner of intermediate conditions. In both rats and humans, maternal style within the normal range can be transmitted transgenerationally.[20] In humans though—in contrast to rats and most other mammals, including primates such as rhesus monkeys and gorillas—fathers, as well as mothers, play an important role in parenting. Paternal style has not been much studied to date, nor its effects on the emotional behavior and stress response of the next generation. But evidence from studies of child abuse and its social transmission suggest an important role for the male parent. Moreover, one recent study found a correlation between levels of the master stress hormone CRH (see Chapter 4) and reported levels of parental, not just maternal care.[21] The social inheritance of paternal style clearly requires investigation.

Bent Twigs

There is much truth to the old adage that as the twig is bent, so the tree inclines. Things that happen early in life have long-lasting consequences. As we have seen, one mechanism behind this tendency is environmentally induced epigenetic change. But in my many walks in the forest, I have noticed that some trees start growing in one direction, then change direction quite dramatically, sometimes up to ninety degrees. This usually occurs when the tree begins life by growing sideways due to some environmental impediment, like a rock or other trees, then makes a turn toward the vertical and the sunlight. There are numerous analogous cases in human psychological development. Many who get off to a bad start make midcourse corrections. Most victims of child abuse do not become child abusers. That cycle can be broken.

Infant-parent interactions are the foundation of the process of socialization but subsequent events, especially interactions with peers, also figure prominently in social and hence emotional development. This is true even in rats. Michael Meaney and his colleagues were able to reverse many of the adverse effects of poor mothering by providing his subjects an enriched social environment after weaning.[22] After time spent with well-adjusted same-sex peers, there were noticeable changes in the stress response. Significantly, this change was accompanied by changes in the methylation of the NGF gene. Though these epigenetic attachments tend to persist, they are not irreversible.

Primates such as rhesus monkeys can also be rehabilitated in similar ways. There are limits, of course. Harlow's motherless monkeys could not be rehabilitated, nor could Charles the gorilla ever be properly socialized beyond sexual intercourse. And Charles was fortunate in that respect, compared with many other motherless male gorillas. Sometimes the twig is bent too much to be remedied.

In humans, given our protracted period of socialization, the opportunities for overcoming an adverse childhood seem especially promising. To the extent that children at risk experience successful remediation, we would expect to find epigenetic reversals of the sort identified by the Meaney lab, as well as perhaps new epigenetic alterations in other genes. (Epigenetic processes do not end or begin with childhood.) For intractable cases where pharmaceutical intervention is deemed necessary, the most effective drugs will be those that epigenetically alter gene expression. The Meaney lab successfully used such pharmaceutical treatments to remedy the stress response of rats that were victims of poor mothering.[23]

Those badly bent twigs that are not remedied by subsequent alterations to their social environment develop into prostrate trees that bear poor fruit. This is the pathological dimension of social inheritance, the dimension of social inheritance that is easiest to identify, but only the tiny, exposed portion of the iceberg.

Expanding Our Notion of Inheritance

We inherit more than our genes. Among the extragenetic things we inherit is a social environment that begins with our parents but can extend well beyond that, up to and including a whole culture. Gorillas inherit their social environment as well. Charles and other captive gorillas dramatically evidence what can go wrong in the socialization process when this social environment is aberrant. Harlow created even more extreme pathological social conditions in his maternal deprivation experiments. His motherless mothers could not begin to adequately care for their offspring. Charles, at least, had bonded with his mother before the poachers killed her. Many captive-born gorillas aren't so lucky. While not subjected to the extreme deprivation of Harlow's monkeys, they are raised by poor gorilla-mother substitutes called humans, and become inept fathers or neglectful mothers as a result, in a perpetual cycle of pathological socialization.

Social inheritance of a less pathological sort was long ago demonstrated in rats. A female rat subjected to stress transmits her elevated stress response to her daughters and her daughters' daughters. But even unmanipulated rats that exhibit the normal range of maternal behavior (as evidenced by high or low grooming rates) tend to transmit their maternal style to their female offspring. In this case, though, scientists have uncovered the mechanism, which involves epigenetic changes in two genes: NGF and an estrogen receptor. It will be worth exploring whether the same epigenetic processes underlie the transmission of maternal styles in nonhuman primates and humans. In humans, unlike in most mammals, the transmission of paternal style is worth exploring as well.

The Meaney lab has shown that the effects of poor parenting can be reversed through an enriched social environment for the offspring of rats within the normal range of maternal behavior. Not surprisingly, the behavioral changes are associated with epigenetic changes in the NGF gene. The pathologies induced by Harlow on his rhesus monkeys and by captivity on gorillas are more refractive. But though human child abuse runs in families, most of those abused do not become abusers, suggesting an important role for subsequent socialization.

Though parenting style, including child abuse, is not transmitted from one generation to another with the same fidelity as classical genetic inheritance of, say, eye color, it is an important form of inheritance nonetheless, especially with regard to psychosocial behavior. In fact, social inheritance can be easily mistaken for classical genetic inheritance involving "abuse genes" and such. Genes are involved, to be sure, but as effects rather than causes. To the extent that genes play a causal role in the social inheritance described here, it is through their environmentally induced epigenetic modifications. Is this, then, a form of epigenetic inheritance? A case could be made that it is an indirect form for epigenetic inheritance. The difference between this indirect epigenetic inheritance and direct epigenetic inheritance will be explored in the next chapter.

Chapter 7

What Wright Wrought

WHEN YOU THINK OF DOMESTICATED ANIMALS, THE GUINEA PIG is not one of the first that comes to mind. Yet the guinea pig was domesticated a thousand years before the horse—not as a pet but for food. In fact, to this day, guinea pigs remain a dietary staple in the Andes of Peru and Bolivia, where they were originally domesticated. It was only thousands of years later, after they were transported to Europe during the seventeenth century, that they were made into pets, and later the paradigmatic subjects for scientific experiments.

It was in part because of their circuitous route—by ship—to European ports that guinea pigs acquired their common moniker. And a curious name it is. The guinea pig is neither from Guinea, nor a pig. The "guinea" in "guinea pig" refers to the Guinea coast of West Africa, a stopover and resupply point for European ships heading to and from South America. Though the Guinea pigs were originally loaded on the ships in South America, many Europeans mistakenly assumed that they had been supplied in Guinea.

European sailors, like the Andean people who domesticated them, loaded their ships with guinea pigs because they were a good source of protein for the long journey. Some of those that didn't get eaten became the first guinea pig pets.

The "pig" in "guinea pig" probably derives in part from the fact that they didn't look anything like European rodents—mice, rats, and such—though their resemblance to pigs is not all that evident either. But so they seemed at least to the father of scientific classification, Linnaeus, who baptized them with the species name *porcellus* (the Latin word from which "pork" is derived), perhaps because their bodies are fairly squat and their tails tiny.

The guinea pig is actually a type of cavy, a group of South American rodent species that generally inhabit high-altitude grasslands. The bulk of a wild cavy's diet is grass; indeed, they are considered by some the ecological equivalent of cows. But guinea pigs are not as grass bound as cows; they can thrive on a diverse diet, which is one reason they are easy to maintain in captivity. In the Andes they often have the run of the house until the dinner bell tolls.

In the eyes of well-fed Europeans, guinea pigs seemed irresistibly cute and cuddly. Hence their conversion in Europe from food to pets. During this second round of domestication, guinea pig fanciers selectively bred for a variety of coat colors as well as for long and curly hair, which diverged quite dramatically from the wild type. It was primarily because of this variation in pelage that guinea pigs became the first mammalian model for genetic studies.

William Castle and Sewall Wright

At about the same time that Morgan was setting up his Fly Room at Columbia University (see Chapter 2), William Castle set out to do comparable genetic research at Harvard. Somewhat ironically, it was Castle who first saw in fruit flies their potential value for genetics,[1] but unlike Morgan he chose to stick with mammals. Castle had labs devoted to rabbits, mice, and rats, but he had a special fondness for guinea pigs. He became so enamored that he traveled to South America to collect their wild relatives for his breeding experiments. Guinea pigs are far from ideal subjects for genetic studies, though, far less suitable certainly than fruit flies. Fruit flies can cycle through fifty generations in a year, guinea pigs two, maybe three, if you are lucky. Morgan's group had identified over one thousand mutations before Castle's group found ten. Yet guinea pigs have some advantages over fruit flies, if only that any mutation is easier to spot. Mutations in coat characteristics are especially easy to see.

Castle had a hands-off managerial style similar to Morgan's. He gave his students great leeway in selecting their projects. And though Castle's lab was not as teeming with future luminaries, it did produce several who became members of the National Academy of Sciences, including one very special student named Sewall Wright, whose contributions to genetics were unsurpassed in their breadth.[2] Wright not only greatly enlarged the scope of classical genetics as it relates to heredity; he also thought carefully about the gene as a physiological element and how a gene could developmentally influence a trait like coat color. For this reason, he can be considered one of the fathers of developmental genetics. But Wright is best known as one of the co-founders of the field of population genetics, through which he had a huge influence on evolutionary biology. Wright had a unique set of aptitudes and interests, which made all this possible.

Wright, who was something of an autodidact, also came from a different biological tradition than Morgan and his students: physiology. One consequence of his background was his concern with what genes actually do physiologically. Of course, in this premolecular age, in which the physical gene had not yet been characterized, such inferences were extremely indirect. Wright nonetheless was remarkably prescient as to the physiological actions of the genes he studied. Moreover, because of his concerns about the concrete physiological gene, as opposed to the Mendel-Morgan notion of genes as abstract particles of inheritance, Wright was less inclined to try to shoehorn seemingly anomalous results into the Mendelian framework.

In many breeding experiments conducted during this period, there would be slight to sizeable departures from what would be expected from Mendel's laws. The sizeable departures were taken seriously, but lesser departures were generally considered within the margin for error. Wright, though, chose to emphasize this residual variation rather than ignore it. Ultimately, this tactic caused Wright's view of genes and gene actions to diverge substantially from the mainstream of classical genetics. In particular, he tended to a more complex view of genetic inheritance than was the norm of the time. First, Wright came to believe that the inheritance of traits such as coat color involved more loci and alleles than most geneticists. Second, he emphasized the fact that alleles at different loci can interact in complex ways that cannot be calculated from their separate effects, a phenomenon known as *epistasis*.

Though now widely accepted, this view was not at all well received by his contemporaries. Wright had other ahead-of-his-time ideas as well. For example, Wright was much more open to the possibility that the environment contributed to coat color and other "genetically determined" traits, and was a pioneer in the study of gene-environment interaction. This, too, ultimately stemmed from his being steeped in the physiological or developmental view of genes and gene actions. Because Wright viewed genes as physiological entities, it was much easier for him to envision that the effect of a given gene can depend on environmental factors that also influence the same physiological processes.

But perhaps most heretical was the importance that Wright ascribed to random events in modulating gene effects and development. Wright saw randomness at every level of biological processes, from the biochemical to that of the whole guinea pig, and beyond that, to whole populations of guinea pigs.

Wright's somewhat iconoclastic brand of (physiological) genetics was always overshadowed by the classical school, which led from Morgan to Watson and Crick, but it was Wright's approach to genetics, not Morgan's, that eventually laid the foundation for epigenetics. The study of epigenetic inheritance, the subject of this chapter, certainly owes much more to Wright than it does to Morgan.

The Agouti Gene

By the time Wright commenced his research on guinea pigs, several genes were known to contribute to coat color. Wright devoted much of his early career to figuring out how they interact to produce the various color patterns guinea pig fanciers had created, plus some additional ones that Castle had created by hybridizing domesticated guinea pigs with their wild progenitors. Wright began with the Mendelian assumption that each color-related gene (locus) acted independently of the others and that there were two variant alleles at each locus: the wild-type allele, and a mutant allele made more common through the selective breeding of guinea pig fanciers. He further assumed, like Morgan, that the wild-type alleles were dominant and the new mutant alleles recessive. These Mendelian assumptions served Wright well for the most part, but there was always a substantial amount of residual variation that could not be explained within the standard Mendelian framework.

Wright's research on the agouti locus is particularly germane for our purposes here.[3] The agouti locus is named for the color pattern displayed by agoutis, which are basically long-legged versions of guinea pigs. The wild-type agouti allele, call it *A*, is usually associated with a distinctive color pattern. Each hair starts out black; that is, it has a black tip. As the hair continues to grow, however, it becomes yellowish to red, then black again at the base. These banded hairs, it turns out, are typical of most wild mammals, not just agoutis, including the wild progenitor of guinea pigs. And the agouti gene is found in all mammals including humans.

Breeders had subsequently selected for a mutant allele, *a*, that caused the width of the yellow band to increase at the expense of the black, resulting in various reddish-yellow color patterns. But these two alleles could not account for all of the variation in hair banding. Wright demonstrated that there was at least one more genetic factor involved, a second mutant allele at the agouti locus. But even given this third allele, there was still unexplained variation. Wright was certainly willing to acknowledge that environmental factors might be complicating things, but given the state of the art of that time, he could not have envisioned how. He would not, however, have been surprised at the answer. It falls quite neatly into his worldview.

Though Wright continued to use guinea pigs for his genetic studies, most subsequent research at the agouti locus was conducted on mice. Wright's original research on the agouti locus in the guinea pig laid the groundwork for the research on this same gene in mice, including the recent epigenetic research that we will explore here.

In the nineteenth century, the pet trade in mice rivaled that of guinea pigs. Mouse fanciers had also uncovered recessive mutations in the agouti locus that caused the yellow band to widen, resulting in a yellowish coloration. But the number of mutations at the agouti locus increased considerably when scientists took over the breeding. Moreover, some of these mutations were dominant to the wild-type allele, A. One such mutation, *lethal yellow* (A^L), was generally lethal. Other dominant mutations were not lethal. These included the mutation *viable yellow* (A^{vy}), so called because, unlike mice with the lethal mutation, mice with this mutation survived, albeit with significant physiological defects. *Pleiotropic* is the term used to describe alleles such as A^{vy} that have multiple physiological effects.

Pleiotropy simply reflects the fact that the protein products of most genes are expressed in more than one cell type. As such, a gene can participate in more than one physiological or developmental process. In this case, the most obvious developmental process—to the human eye—that the agouti gene participates in is the one that determines hair color. The agouti protein affects hair color by interfering with the binding of the hormone that promotes melanin (associated with black pigment) production to its receptor.[4] But melanin is produced in many cell types other than hair follicles; the agouti protein interferes with melanin production in all of them, including those found in the liver, kidney, gonads, and fat.[5] The result of all this interference is lethal in A^L (*lethal yellow*) mice and grossly compromises health in A^{vy} (*viable yellow*) mice. Among the adverse health consequences of this mutation are obesity, diabetes, and various kinds of cancer.[6]

Unlike A^L mice, which are always yellow, the coat color of A^{vy}

mice is quite variable, ranging from almost pure yellow to a wild-type coloration, called pseudoagouti. You can predict the health of a viable yellow mouse by its coat color. Those viable yellow mice with yellow coloration are obese, diabetic, and cancerous; those with the pseudoagouti coloration, however, have none of these defects.[7]

How do we explain this variable coloration and associated health defects of viable yellow mice? One explanation, consistent with Wright's approach, concerns what has come to be called *genetic background*. For Wright, in contrast to most of his contemporaries, the effect of a gene (allele) like *viable yellow* on a trait such as coat color depends on a lot of other factors. Among these other factors are other genes. That is, the effect of the *viable yellow* allele on coloration depends, in part, on what other alleles are present at other genetic loci. Not all other loci, of course, but a lot more loci and alleles than most of his contemporaries were willing to acknowledge.

But the *viable yellow* allele has variable effects on coloration and health even when the genetic background is kept constant. That is, even genetically identical mice with this mutant allele are quite variable in coat coloration and health. Within a single litter of genetically identical *viable yellow* mice, you can find the full range of coat patterns from yellow, to mottled, to pseudoagouti, and with them the associated variation in health.[8]

Epigenetics at the Agouti Locus

The color differences in these mice, it turns out, are associated with differences in the epigenetic state of the *viable yellow* allele. It is unmethylated in the yellow mice but highly methylated in the pseudoagouti mice. Mottled mice are intermediate methylation-wise.[9]

So why do some of these genetically identical *viable yellow* mice have methylated *viable yellow* alleles, while others don't? In part, it depends on the coat color, and hence epigenetic state, of the mother. Female mice with the yellow coat color tend to produce yellow offspring, never offspring with the pseudoagouti phenotype. Mothers with the pseudoagouti coloration produce few yellow offspring and more pseudoagoutis.[10] Moreover, a grandmother's coloration also influences the coat coloration of her grand-offspring.[11] There is no relationship between the coat color of a father and that of his offspring.

This may sound familiar, like the transgenerational maternal effects on the stress response that we discussed in the previous chapter. But though these curious patterns of coat color inheritance certainly constitute a maternal effect, this one occurs much earlier in development. When the fertilized eggs of yellow mothers were transferred to black mothers, they still tended to be yellow at birth.[12] So there is no effect of the intrauterine environment here. Instead, an epigenetic attachment to the A^{vy} allele, one that alters coat color in otherwise genetically identical mice, has been directly transmitted from mother to pup. This is true epigenetic inheritance.

But how did this epigenetic variation arise in the first place? From previous chapters, we might expect some sort of environmental effect. In this case, diet appears to have a role. When pregnant *viable yellow* mothers are fed a diet high in methyl donors, such as folic acid, the spectrum of coat colors in their offspring shifts toward the pseudoagouti end.[13] Moreover, when the offspring, which experienced the methyl supplementation in utero, themselves became mothers, the color spectrum shift is sustained in their offspring.[14] The transmission of the diet-induced change to the grand-offspring occurred even though these second-generation mothers received no further methyl supplementation.

The shift in the color spectrum caused by a high methyl diet was quite modest. Most of the epigenetic and hence color differences in genetically identical *viable yellow* mice must result from other factors. One source for this variation, which is receiving increasing attention of late, is chance. Much of what causes one individual with this allele to be yellow or pseudoagouti—with all of the associated health consequences—may be essentially random processes at the molecular level that affect methylation of the allele.[15] So in essence, we have a case of a partly random epigenetic event that can be inherited. That sounds very much like a mutation.

Why Epigenetic Inheritance Was Not Supposed to Occur

For years, it was thought that true epigenetic inheritance was impossible. During the process of making sperm and eggs, all epigenetic marks were thought to be removed during a process called *epigenetic reprogramming*.[16] Any epigenetic attachments that survived this process are removed during a new round of reprogramming soon after fertilization. Hence each new generation starts with a clean epigenetic slate. Recently, however, it has been demonstrated that epigenetic reprogramming does not wipe out every epigenetic mark. Some epigenetic changes, including those induced by environmental factors, are not erased; they are transmitted to the next generation.

The agouti locus is one of the best-documented cases of epigenetic inheritance among mice, but there are a number of other known cases in mice as well. One such case involves the *Axin* gene, which when methylated results in a kinked tail.[17] The methylation pattern and hence kinked tail can be inherited from both mother and father. A number of genes related to olfaction, especially the detection of pheromones, also display putative epigenetic inheritance.[18] In humans, there may be epigenetic inheritance at a locus that promotes a particular form of colon cancer.[19] Since it is only recently that cases of anomalous (by Mendelian standards) inheritance have been viewed through an epigenetic lens, we might expect more cases of epigenetic inheritance in humans and other mammals will be discovered in the near future.

But there are reasons to suspect that epigenetic inheritance is less common in mammals than in other life-forms.[20] Good examples of epigenetic inheritance have been identified in creatures as diverse as fruit flies and yeast.[21] But some of the most dramatic examples of epigenetic inheritance occur in plants.[22]

As an experimental subject, the plant equivalent of a mouse is an unprepossessing member of the mustard family known only by its scientific name, *Arabidopsis thalinia*. In the wild, *Arabidopsis* thrives in diverse habitats throughout Eurasia; it also thrives in laboratory environments. It is a highly variable plant with respect to its size and flowering time, among other traits. Both the size of the plant and its flowering time are inherited epigenetically. Let's consider first an epigenetic factor that affects the size of the plant.

In many organisms, but especially plants, there is a tradeoff between growth and defense against infection by pathogens. The more resources devoted to pathogen defense, the slower the growth rate. So plants that find themselves in a particularly pathogen-rich environment tend to be dwarfs. Pathogen resistance in *A. thalinia* is mediated by a number of resistance (R) genes. There is one particular cluster, located on chromosome 4, that is subject to epigenetic regulation. An epigenetic variant called *bal* causes one gene in this cluster to be chronically active. The gene behaves like the plant is under attack even when it isn't. Plants with this epigenetic variant are scrawny and bedraggled looking, their leaves withered, and their roots underdeveloped. But genetically identical plants that lack this epigenetic factor are robust, even when they are grown in identical environments.[23] Before the advent of epigenetics, it would have been assumed that the dwarf was a mutant, that there was some change in the sequence of the R gene. Now we know that even such stark differences within a species can be caused by differences in the epigenetic regulation of gene expression.

Flowering in *A. thalinia* is also epigenetically regulated. In 1990, a mutation was identified in some wild populations of *Arabidopsis* that caused a delay in flowering.[24] The mutant, called *fwa*, caused plants that would normally flower in the spring to flower instead during the summer or fall. Multiple genetic tests indicated that *fwa* was a classical dominant Mendelian trait, like brown eyes in humans. The mutation was subsequently mapped to a single gene encoding a transcription factor. But scientists were puzzled when they could not find any difference in the sequence of the *fwa* mutant and the normal *FWA* allele. The mutation, it turned out, was not a mutation but an *epimutation*, an altered methylation pattern. And this epimutation has been stable, inherited in a quasi-Mendelian manner, for many years.[25]

Transgenerational Epigenetic Effects

Epigenetic inheritance, such as occurs at the agouti locus and in *A. thalinia*, is but one form of what I will call a "transgenerational epigenetic effect," by which I mean an epigenetic effect transmitted from parent to offspring and beyond.[26] This broader category includes the social inheritance of the stress response in mice and other forms of nongenetic inheritance that have an epigenetic component. To qualify as epigenetic inheritance in the strict sense, though, the epigenetic attachment, or mark, must pass through the epigenetic reprogramming process intact. In the case of the lick-deprived rats' stress response, the epigenetic attachments are reconstructed anew each generation; the original epigenetic alterations do not survive epigenetic reprogramming. This is probably true of most transgenerational epigenetic effects that result from the maternal or social environment. That would include the effects of the Dutch famine discussed in Chapter 1. There is no compelling evidence that the grandmother effect described there is true epigenetic inheritance. But in another study of dietary effects in humans, the case for true epigenetic inheritance is much stronger.

There is an isolated Swedish population for which very accurate records of crop harvests have been kept for hundreds of years, so scientists can calculate the average amount of calories consumed in a given year. One notable result of this study is an association between the calories consumed by men during their adolescent years and the health of their grandchildren. Paternal (but not maternal) grandsons of males exposed to famine before adolescence were less susceptible to cardiovascular disease than the grandsons of those who did not experience famine.[27] In contrast to the epigenetic effects on birth weight resulting from the Dutch famine, this association cannot be attributed to the maternal environment. The only biological thing a male contributes to his offspring and grand-offspring is his sperm. So this appears to be a case of environmentally induced epigenetic changes that qualify as epigenetically inherited.

It should be noted that epigenetic inheritance at the agouti locus is not very precise or efficient. The correlation between parent and offspring, while significant, is not high. This parent-offspring correlation is in fact much lower than that for the stress response in rats, which does not result from true epigenetic inheritance.

It is different in plants. True epigenetic inheritance is much more common in plants than in animals and can be stable over hundreds of generations—as stable, in some cases, as genetic inheritance. The reason for the higher incidence of epigenetic inheritance in plants is that their epigenetic reprogramming is much less pervasive and thorough. Many more epigenetic marks make it through this process unscathed.

Wright's Legacy

Sewall Wright's work on the inheritance of coat color in guinea pigs, which continued throughout his long life, is a testament to both his close observation and a theoretical insight that was informed, but not encumbered, by established doctrine. Wright's approach to genetics differed markedly from that of most of his contemporaries in its emphasis on the gene as a physiological and developmental factor. Most geneticists of his time, including Morgan and his group, preferred to view the gene as an abstract hereditary factor. The payoffs for Morgan's approach were immediate, whereas most of the dividends resulting from Wright's approach would become apparent only decades later. It was Wright's approach, however, not Morgan's, which laid the groundwork for developmental genetics and its offshoot, epigenetics.

More specifically, Wright's research on the agouti locus became the foundation for subsequent research on its developmental role, from hair color to obesity. Ultimately the ongoing research on the agouti locus resulted in the first well-documented case of true epigenetic inheritance in a mammal.

As mentioned earlier, epigenetic inheritance in the strict sense is but one form of the broader phenomenon of transgenerational epigenetic effects, other forms of which—such as the social transmission of the stress response in rats—we explored earlier. We will explore another kind of transgenerational epigenetic effect in Chapter 9, a rather odd one; but first some background information will be helpful, which comes to us by way of the mysterious X chromosome.

Chapter 8

X-Women

AS A CHILD, I ALWAYS PREFERRED GAMES INVOLVING PHYSICAL activity to board games. In part, this was because I found board games boring; in part, because my sister was clearly superior. Monopoly was particularly unpleasant in both regards. But a broken leg at age eight severely restricted my physical activities. Still untempted by board games, I focused much of my attention on a game called Caroms, a sort of billiardish affair played on a small, square wooden surface, with wooden sides and four corner pockets. The wooden cues were about eighteen inches long. The object was to pocket the small doughnut-like objects, also made of wood, which came in two colors: red and green. You were responsible for knocking in all of the reds, or greens, as the case may be—like the solids and stripes in pool. Whoever knocks in all of the appropriate wooden disks wins.

Caroms had at least two virtues from my perspective: it required some physical skills, and I completely trounced my sister at it. In fact, I pretty much beat everyone. Most of my friends presented a bigger challenge than my sister, who in any case soon developed an aversion to the game. But one of my friends, Steve, was even worse than my sister, for reasons that I initially could not comprehend. It certainly wasn't for lack of hand-eye coordination. Steve was fine in that regard. The problem was that Steve was quite indiscriminate as to which caroms he pocketed. At first I attributed this to boredom, as Steve was even more physically active than me. I thought he might just be trying to get the game over with. But he actually seemed to enjoy it, and when I called his attention to the fact that he had just pocketed one of mine, he would just smile. I found his smile disconcerting, because Steve was much more worldly than I; it was he who first informed me that there was no Santa Claus and that the Tooth Fairy was in fact my mom. So initially, I suspected that he had a motive for "losing" that was unavailable to the uninitiated. This greatly reduced any pleasure I derived from victory. Eventually, I grew so frustrated with Steve that, when he was obviously aiming at a carom of mine, I would call his attention to the fact. Steve would just smile and proceed to knock it in anyway.

At some point, I got my mother involved in order to get to the bottom of Steve's seeming perversity. It didn't take her long to discover that Steve was color-blind. He was pocketing my reds with his greens because he couldn't see the difference. Steve was aware of this at some level, but he was too proud to admit it, hence his disconcerting smile. The fact of Steve's color blindness did not jar my worldview quite as much as the fictitiousness of Santa Claus, but it did initiate some protophilosophical reflections in my young mind. What does the world look like to Steve—flowers? trees? traffic lights? Especially traffic lights. How would he know when to cross the street when he was alone? Color blindness was fascinating stuff.

Over the years, I have found myself returning at odd intervals to the subject of color blindness, most recently because of an epigenetic connection, the subject of this chapter.

Why Males Are Truly the Weaker Sex

Let's begin with the fact that the red-green color blindness that Steve exhibited is much more common in boys than girls. In that respect, it resembles a host of other developmental defects, from dyslexia to certain forms of heart disease. Such defects are said to be sex-linked. Sex linkage occurs when the relevant genes reside on the sex chromosomes, overwhelmingly, the X chromosome. The X chromosome is the largest and most gene-rich of our chromosomes, so lots of our traits reflect some degree of sex linkage—or X linkage, to be more precise. The Y chromosome, on the other hand, is a tiny little thing.

Sex-linked mutations have a signature pattern of inheritance. This is particularly true of recessive mutations—those that must be present on both chromosomes—the one you got from your mother and the equivalent one that you got from your father—to have any effect.[1] To varying extents, this pattern applies to genes on all other chromosomes, collectively referred to as *autosomes*, but not to the sex chromosomes—at least, not the male sex chromosomes. Females are blessed with two X chromosomes, one from each parent. Males, on the other hand, only inherit one X chromosome—from the mother—along with the diminutive Y chromosome from the father. As such, any recessive mutation on the maternally derived X chromosome is effectively dominant in males and causes problems. Hence, males are affected at a much higher rate by these recessive mutations than females. The male X deficit is certainly part of the explanation for the fact that at any age or developmental stage, from before birth to dotage, a male has a higher risk of mortality than a female.[2]

Among the many genes on the X chromosome are two that specify opsins, the color-sensitive proteins in cone cells, our color detectors in the retina. There is a third opsin gene, but it resides on chromosome 7, not on the X chromosome.[3] Since only one opsin gene is expressed per cone cell, there are three distinct types of cone cells, which we can call red, green, and blue cones. The red and green opsin genes reside on the X chromosome, the blue opsin on chromosome 7. When inherited in a male such as Steve, a recessive mutation in either the red or green opsin gene results in defective red or green cone cells, and hence in red-green color blindness. But even if Steve's sister had inherited the same mutation from her mother, she would not be color-blind unless she also inherited the mutation on the X chromosome bequeathed by her father, in which case her father must have been color-blind.

That, at least, is the standard textbook explanation for sex-linked traits that I learned in Introductory Genetics. But there must be more to this sex difference, if only for this startling fact: some female carriers of these mutations actually have enhanced color vision.[4] These mutant women see color differences that no mortal man can. Let's call them X-women.

In this chapter, we will get to the bottom of the X-women phenomenon. This will require that we explore a new epigenetic mechanism, one that involves a high degree of randomness. It is appropriate that we use the X chromosome in this regard, because it was in exploring its mysteries that much of the foundation for the science of epigenetics was laid.[5]

A Dosage Problem

As bad as things are for males X-wise, they would be a lot worse were it not for a process called dosage compensation, which helps level the physiological playing field. Without dosage compensation females would have twice as much of every X chromosome–derived protein as males. This would require that the characteristics of male and female diverge well beyond even those imagined by the most dyed-in-the-wool evolutionary psychologists. And males would be downright feeble compared with females. (Think of those deep-sea anglerfish in which the tiny male attaches to the first giant female that happens by, at which point he degenerates into a physiologically parasitic sperm purveyor: a wartlike testis and little else.)

The evolutionary solution to this dosage problem is called X inactivation,[6] in which one of the two X chromosomes in every female cell is inactivated. As a result of X inactivation in females, both males and females have one functional X chromosome per cell. But if both males and females are working with one functional X chromosome, why do males have so many more X-linked problems than females? It turns out that even though females are essentially operating with one X chromosome tied behind their backs, they still derive many of the benefits of having two.

Part of the explanation is that not all of the genes on the inactivated X chromosome are inactivated. In humans, 15 to 25 percent of the genes on the inactivated X chromosome escape inactivation.[7] Many of these genes that escape inactivation are referred to as *housekeeping genes,* which participate in basic cellular processes required by all cells, from skin cells to neurons to cone cells.

There is another reason why females have many of the benefits of two X chromosomes despite the fact that one is largely inactivated. In most mammals, including humans, X inactivation is random with respect to the maternal X and the paternal X. And this random inactivation occurs independently in each cell lineage. That means that in a given population of cells—such as, say, red cone cells—roughly half will have paternal X inactivation, and half maternal X inactivation. Females are essentially X chromosome mosaics. If a female inherits a recessive mutation of the red opsin gene, from either father or mother, only half of her cone cells will be affected, whereas all of a male's red cone cells would be affected by any such mutation. Operating with half the amount of normal red cone cells is sufficient to avoid color blindness as defined by standard tests, but as we shall see, there can be subtle deficits in the color perception of such females.

In marsupial mammals (kangaroos, koalas, wombats, and such) X inactivation is not random. Instead, the paternal X is always inactivated.[8] Hence, male and female kangaroos both operate solely with the maternal X, and are therefore physiologically equivalent with respect to the X chromosome.

Epigenetics of X Inactivation

X inactivation is initiated at a site called the X-inactivation center (Xic). There are several genetic elements within Xic, one of which is particularly crucial for X inactivation: X-inactive-specific transcript (Xist). Sometimes bits of one chromosome get dislodged and land on a different chromosome, a process called *translocation*. If the bit of the X chromosome containing Xist is translocated to one of the autosomes in this way, the X chromosome cannot be deactivated. The autosome on the receiving end gets (partially) inactivated instead.[9] So Xist is absolutely essential for X inactivation.

Xist is not actually a gene in the traditional sense of the term. A gene, you will recall, acts as an indirect template for a protein. But there is no Xist protein, only Xist RNA. That is why it is called X-inactive specific transcript, not X-inactive specific protein (or

Xisp). The Xist RNA is quite long and it attaches to the X chromosome from which it is made. As more and more copies of Xist RNA are produced, the X chromosome becomes covered with the stuff, the first stage of inactivation. Next, the Xist RNA attracts histones (see Chapter 5)—which further coat the inactive X—as well as methylating factors. Finally comes the big crunch, when the inactivated X is compacted like a scrapped automobile. Under a microscope, the compacted X chromosome is a little bloblike structure called a Barr body, which doesn't look anything like the active X chromosome. The compacted X chromosome is still much larger than the Y chromosome, however.

Earlier, I stated that X inactivation is random. That is not strictly true, for two reasons. The first has to do with the timing of X inactivation. We do not know, with any precision, when exactly X inactivation occurs during early development, but it occurs long before birth. There are many cell divisions subsequent to X inactivation, and a given cell lineage retains the X-inactivation pattern of its original X-inactivated founder. So it would be more accurate to say that X inactivation is random with respect to cell lineages, a particular population of hair cells, or cone cells, say. This is easiest to discern in certain patterns of hair (coat) coloration in mammals like cats. Calico and tortoiseshell patterns are particularly useful in this regard, since both colorations are X linked and confined to females. A calico cat reveals in great detail the random X inactivation of hair cell lineages, in the distribution of light and dark and orange areas. It is ironic, therefore, that the first cloned cat was a calico. The owner wanted to recreate her beloved cat, Rainbow. The cloning procedure was successful, but the clone, named Cc (for "carbon copy") was not even close to a carbon copy.[10] She developed a completely different distribution of colors than Rainbow. Given the randomness of X inactivation, this should have been expected.

(Cc also displayed a much different personality than Rainbow, but that's a different story.)

X inactivation is also not random with respect to the maternal tissue that sustains the fetus. Instead, only the paternal X chromosome is inactivated, as in kangaroos and other marsupials.[11] The selective inactivation of the X chromosomes from one sex is a form of imprinting, a phenomenon I will discuss in the next chapter. For now it is sufficient to note that in kangaroos, X-chromosome imprinting is pervasive, extending to most cells; for cats and humans, the imprinted X chromosome is confined to cells in the placenta and some other extraembryonic tissue.

The marsupial form of X-inactivation is considered the primitive condition for mammals. The random X inactivation of cats, humans, and other mammals in the more modern mammal lineage evolutionarily diverged from that of the marsupials. The signal event in this regard was the advent of Xist. Marsupials lack Xist and therefore the benefits of random inactivation. Indeed, Xist RNA may constitute the most important single difference between marsupials and more "advanced" mammals like us.[12]

X Inactivation and Cone Cells

Cc, the calico cat clone, is testimony to the fact that, due to random X inactivation, we should expect female clones to be more variable than male clones for any X-linked trait.[13] The greater female variability should extend to nontwins as well. That certainly seems to be the case for color vision. Within the normal range of color vision, that is, excluding the color-blind like Steve, human females are more variable than males in color discrimination tests.[14]

At the low end of the range for normal color vision, some female carriers of the red-green mutation are less sensitive to red-green differences than noncarriers and normal males.[15] We can attribute this to the fact that they have fewer normal cone cells. On the other hand, some females can make finer red-green distinctions than normal males. Paradoxically, these ultrasensitive females may also harbor red-green mutations that cause color blindness in males. These are the ones I refer to as X-women. How do we explain these X-women?

Let's begin with a closer look at normal cone cells. The three types of cone cells are distinguished by the wavelengths of light to which they are most sensitive, which in turn depends on the type of opsin they express.[16] Red cones are sensitive to long wavelengths, green to intermediate wavelengths, and blue to short wavelengths. Color perception occurs when the brain integrates the inputs from these three cone cell types. We will confine ourselves to the red-green (long to intermediate wavelength) part of the spectrum. Normally, red and green opsins have different peak sensitivities with respect to wavelength. The brain therefore receives two distinct inputs. Red-green color blindness occurs when a mutation causes the peak wavelength sensitivities in red and green cone cells to converge, due to a mutation in the red or green opsin. In essence, the red and green cone cells become more similar with respect to the wavelength of light that causes them to fire their signal to the brain. It therefore becomes more difficult to distinguish red from green. This was Steve's condition.

Red-green color blindness is particularly common for two reasons. First, in normal individuals, the peak sensitivities of red and green cones are not that different. Blue cones, on the other hand, have a much different peak sensitivity than green cones, and of course, even more so with respect to red cones. Second, the genes encoding the green and red opsins are arranged in tandem on the X

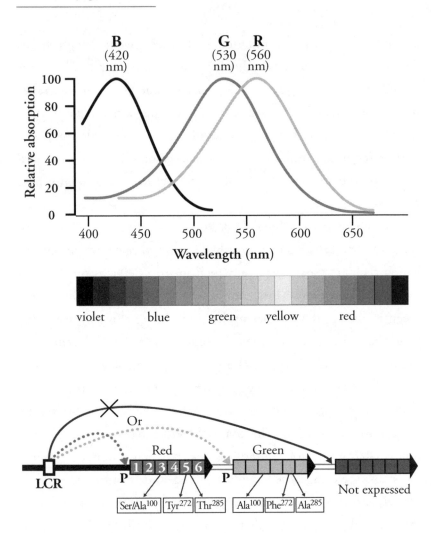

Peak sensitivities for red, green, and blue cones.

Fig. 5 (p. 370) in Deeb, 2005.

chromosome. Adjacent genes are more likely to exchange bits when they are replicated during the process of making sperm and eggs. In this case, such exchanges of gene bits often result in red and green opsins that are more similar to each other than they should be.

But now consider what happens when a similar mutation occurs in a female. Not only is she largely protected from the effects of the mutation because of random X inactivation; she should also end up with four, rather than the typical three, types of cone cells. The blue cone cells, of course, remain unaltered. She also has normal red and green cone cells, albeit half as many as normal. But in addition, she has cone cells with the mutation-induced hybrid opsin. If the peak sensitivity of this hybrid opsin is about halfway between that of normal red and green cones, such a female should theoretically be able to make finer distinctions at the red-green part of the spectrum, and perhaps even between green and blue. Some experiments on color discrimination in such women suggest as much.[17] Ergo X-women.

The X-woman phenomenon has a precedent in other primates. There are two main primate branches on the mammal tree: Old World and New World. Old World primates include African and Asian species such as baboons, macaques, and langurs, along with the great apes and humans (our origin was in Africa); New World primates include spider monkeys, howler monkeys, and capuchins. Color vision in Old World primates, like ourselves, is said to be trichromatic (from the Greek for "three color").[18] We see more than three colors, of course, but all the colors we see are some combination of our three types of cone cells. New World primates, in contrast, have only two types of cone cells, so they are dichromatic.[19] As such, New World primates cannot discriminate colors as well as us Old World primates. But there is a common X-linked mutation in New World primates, which, though it impairs male color vision, provides the females with three functional cone cell types, rendering them trichromatic.[20] The enhanced color vision of these females, again, depends on random X inactivation. A similar mutation in a kangaroo could only impair the color vision of a doe.

An Epigenetic Boon

Color blindness is only one of the ways in which human males suffer disproportionately, for lack of a second X chromosome. Dosage compensation through X inactivation is not sufficient to compensate entirely if it is random, as in humans. Random X inactivation is a boon to females and a huge advance over the kangaroo condition of imprinted X inactivation. Random X inactivation was made possible not by the evolution of a new gene, but rather by a new piece of noncoding DNA that acts as a template for a new kind of RNA, called Xist, which made possible a new form of epigenetic regulation of the X chromosome.

Xist-mediated X inactivation is but one form of RNA-based epigenetic regulation. Most forms of RNA-based gene regulation are more widely distributed throughout plants, animals, and fungi. That is a subject I will take up later. In the next chapter, I will focus on the epigenetic process responsible for the form of X inactivation found in marsupials, for it, too, is an evolutionary advance of sorts. Among vertebrates, it is largely confined to mammals. And this epigenetic process, called imprinting, is not confined to the X chromosome; it occurs throughout our genome, albeit sporadically. Moreover, genomic imprinting occurs even earlier in development than X inactivation does in X-women. In fact, it occurs before sperm meets egg.

Chapter 9

Horses Asses

THERE WERE NO JACKASSES IN THE UNITED STATES WHEN IT WAS founded, yet soon they were everywhere. Where did they all come from?

It goes back all the way to George Washington. Ever curious, Washington kept himself abreast of what was going on elsewhere in the world, particularly as it related to farming. At some point he heard about the amazing exploits of creatures called mules and sought to bring some over from Europe for his own scrutiny. At the time, Spain had a near monopoly on mules, a legacy of the Moors. Actually, there was no monopoly on mules as such; Spain was willing to share mules with the world. The monopoly was on the means of producing mules, a tricky procedure, since mules are not created by conventional means, that is, by other mules. Mules are, rather, the spawn of the unnatural couplings of horses and donkeys. By "unnatural" I mean these things don't just happen when donkeys and horses share a pasture; it takes some coaxing, not least finding two to tango. To increase the odds of success, the human interveners generally seek male donkeys and female horses, because the libido of a male donkey is higher than that of a typical male horse and female horses are less particular than female donkeys—though in their defense, mares are typically blindfolded.

The mules that are the result of this lack of discrimination cannot perpetuate themselves because they are sterile. Their sterility does not deter male mules from attempting to mate; they are as ardent as their donkey fathers but can never be fathers themselves. Mules must be generated anew from further indiscriminate acts between donkeys and horses. So what Washington wanted wasn't a shipload of mules but one of those libidinous male donkeys, so he could make mules of his own. Domestic donkeys are descended from members of the horse family known as asses. For reasons obscured by the veil of time, male asses are called jacks and female asses are called jennies, while all other members of the horse family, including zebras, are called stallions and mares respectively. Hence the term *jackass*, its pejorative connotations deriving from the fact that donkeys of both sexes are less pliable—though by most accounts more intelligent—than horses.

Spain treated their donkeys like the Chinese did silkworms: their export was outlawed. But in 1785, Washington was, by virtue of his prestige, able to persuade King Charles III to part with one jackass named Royal Gift, on which the American mule industry was founded.[1] His unnatural progeny proved invaluable in settling the country, especially as it expanded westward. Mules were particularly valued for hauling and plowing, activities for which they were favored over horses because of their superior strength and sure-footedness. Despite these virtues, mules are memorialized in this country primarily for their stubbornness and ornery disposition. According to William Faulkner, a mule will "work patiently for you for ten years for the chance to kick you once."[2] Elsewhere, however, mules have been known for their physical feats—not their behavioral drawbacks—dating back to the time they were first created over three thousand years ago in the Middle East, where there was a plentiful supply of both asses and horses.

The early mule breeders also sometimes mated horse stallions with jennies, the progeny of which are called hinnies. And from these early days, significant differences between mules and hinnies were noticed. Mules are larger and stronger than hinnies and possess larger donkey-like ears. Indeed, mules look like oversized donkeys with longer legs. Hinnies, on the other hand, are much more horselike in appearance and more tractable. (Hinnies, not mules, are deployed at Disneyland to pull carriages, for example.)

The mule-hinny divide is a 3,000-year-old puzzle, only recently solved. In the process of solving this puzzle, scientists have uncovered a novel transgenerational epigenetic effect called genomic imprinting.

It Depends on Whether It Came from Mom or Dad

The mule-hinny puzzle boils down to this: Both are half-horse and half-mule, so why should they be so different? It violates one of Mendel's fundamental laws of inheritance. We were taught in high school biology that, aside from the Y chromosome, each parent bestows on us separate but equal genetic complements. From each parent, we get one set of chromosomes and hence genes; inheritance is sexually symmetrical. The mule was the first evidence that, in addition to the Y chromosome, there is something else asymmetrical in what is bestowed on us by our mothers and fathers.

This asymmetry came to be known as the *parent-of-origin effect*. It was most obvious in hybrids like the mule. For example, tigons (the progeny of male tigers and female lions) and ligers (male lions and female tigers) are also quite different animals.

While most obvious in hybrids, the parent-of-origin effect has been identified through a variety of other means as well. In humans, Turner syndrome provides a case in point. Turner syndrome is a condition that results when part or all of one X chromosome is missing. Normal females, as we saw in Chapter 8, inherit one X chromosome from each parent. Turner females lack one of the X chromosomes, so their complement of sex chromosomes is denoted XO. Given the discussion about random X-chromosome inactivation in Chapter 8, you might not expect this to be a problem. Normal XX females, after all, have one functional X chromosome per cell. But recall also that the entire X chromosome is not inactivated; some genes normally escape inactivation.

Many problems of XO females can be traced to the 15 percent of the genes on the X chromosome that normally escape X inactivation. In normal XX females, both the maternal and paternal copies of these genes are expressed in all cells. In XO females, only one copy is available. This is probably why 98 percent of Turner females are spontaneously aborted. Nonetheless, Turner syndrome is present in one out of every 2,500 female live births, making it one of the more common major genetic defects.

Turner females who survive birth are subject to a number of maladies, most characteristically a failure to sexually mature. Other problems associated with the syndrome, to varying degrees, are poor growth, cardiovascular disease, osteoporosis, diabetes, and problems with spatial cognition.[3] Which of these ailments a given Turner female has depends in part on whether the remaining X chromosome came from her mother or her father.[4]

Turner syndrome provides only limited insight into parent-of-origin effects because so much of the genome is missing. Prader-Willi syndrome (PWS) is more useful in this respect. PWS, too, is associated with a host of developmental abnormalities; these typically include obesity, poor muscle tone, undeveloped gonads, small stature, and cognitive deficits.[5] There is more than one way to generate PWS, but most who suffer this malady have lost a small chunk of chromosome 15, which geneticists call a *deletion*.[6] Within the deleted sequence are several genes and nongenic sequences (DNA that isn't part of a gene). It is not surprising that a deletion of this sort would significantly affect development; what is surprising, though, is that this deletion results in PWS only when it is inherited from the father. If the same deletion on chromosome 15 is inherited from the mother, a completely different disorder, known as Angelman syndrome (AS), results.[7] It's as if the mother's genes in this region have different *stamps* than the fathers' genes. Both maternal and paternal stamps are required for normal development.

It is the providence of these stamps and not simply being double stamped that matters, as demonstrated by those cases of PWS in which there are no deletions. About 25 percent of PWS cases result from a different kind of molecular screwup, in which two copies of the maternal chromosome are produced, rather than the normal condition of one maternal chromosome and one paternal chromosome.[8] In these cases, it is especially clear that a paternal stamp on certain genes located on chromosome 15 is required for normal development.

Some of the relevant genes in PWS/AS have been identified, but it is from studies of a third disorder, Beckwith-Wiedemann syndrome (BWS), that a particular stamped gene has been best characterized. As it happens, this gene is IGF2, which we first met in discussing the Dutch famine (Chapter 1). Recall that IGF2 is a growth factor that is especially important during fetal development.

When *IGF2* has the paternal stamp, it is active; when it has the maternal stamp, it is inactive. This is the normal condition. It is noteworthy that a gene for a protein that inhibits IGF2 actions is also stamped. But the situation is reversed stamp-wise. When the IGF2 inhibitor has the maternal stamp, it is active; when it has the paternal stamp, it is inactive. This, too, is the normal condition.[9] When these parent-of-origin stamps are missing, bad things can happen, one of which is BWS.

Beckwith-Wiedemann syndrome is a growth disorder resulting in an overgrowth of the fetus. BWS is also associated with several other traits, including an increased risk for a particular kind of kidney cancer, called Wilms' tumor.[10] It occurs when either *IGF2* or its inhibitor is improperly stamped. But what is the nature of these stamps? And how are they established?

Parent-of-Origin Stamps Equal Genomic Imprinting

For most genes, both the version (that is, the allele) inherited from the mother and the one inherited from the father are expressed, when they are expressed at all. This typical condition is called *biparental* expression. For about 1 percent of our genes, however, only one of the two alleles is normally expressed. Sometimes it's the allele inherited from the mother; sometimes it's the allele inherited from the father. This is called *uniparental* expression. Uniparental gene expression occurs when either the maternal or paternal gene is more or less permanently disabled. This disabling process, formerly known as "genetic imprinting," is now called *genomic imprinting*.[11] Imprinting is an epigenetic process in which methylation figures prominently.

But the imprinting process is distinctive in several respects. First, there is the timing. As we saw in Chapter 7, most epigenetic alterations are removed during the process of making eggs and sperm. Imprinted genes are no exception; the epigenetic imprints are erased early in sperm and egg development. But there is a second stage of reprogramming in these reproductive cells. In this second stage, the imprinted methylation patterns are restored to the sperm and egg prior to their maturation, and hence are present during fertilization.[12]

The imprinted gene still must survive a second round of reprogramming, a global demethylation that occurs between fertilization and implantation.[13] Imprinted genes are special in that they don't become completely demethylated during this second round of reprogramming. Other epigenetic processes prevent this from happening. So by the time the embryo implants, imprinted genes are already epigenetically fixed in their expression pattern. This is a good thing because imprinted genes generally do most of their work early in development, long before birth.[14]

The reason the term *genomic imprinting* has come to replace "genetic imprinting" is that the imprint is neither on the gene itself, nor its control panel, nor even on a stretch of DNA adjacent to the gene. Instead, the methylation imprint can be located quite a distance away from the gene whose expression it controls, in what are called *imprinting control regions* (ICRs).[15] In the Prader-Willi case, the ICR epigenetically regulates a number of genes on chromosome 15. Genomic imprinting shares with X inactivation this "remote control" of numerous genes.

There is another epigenetic oddity about imprinted genes: the methylation, which is the imprint, does not always block a gene's expression; sometimes it even enhances it. So uniparental expression could arise because one allele is imprinted "on" or because the other allele is imprinted "off." In what follows, I will simply refer to the "active allele" and the "inactive allele." The imprinted *IGF2* allele is normally active only when inherited from the father. The imprinted allele for the IGF2 inhibitor is active only when inherited from the mother.

The Role of Imprinted Genes in Development

The majority of active imprinted alleles are maternal in origin. Many of these genes are expressed in the placenta and put a brake on embryonic growth.[16] Many of the paternally imprinted genes, on the other hand, seem to promote embryonic growth.[17] In the rare cases in which all paternal imprints are lost, the placenta is undeveloped. Conversely, when all maternal imprints are lost, the placenta is unnaturally large. The imprinting of *IGF2* and its inhibitor illustrates this contrast in microcosm. When *IGF2* is improperly imprinted, such that it is expressed in both alleles rather than one, the fetus experiences the overgrowth characteristic of BWS. The overgrowth is especially pronounced if the maternally imprinted inhibitor is not present.[18] Both occur when there is a duplication of the paternal portion of the chromosome and consequent loss of the maternal parts.[19]

Imprinting malfunctions that cause underexpression of the paternally imprinted *IGF2* allele and/or overexpression of the gene for the IGF2 inhibitor result in growth retardation, as in Silver-

Russell syndrome.[20] So the paternally imprinted *IGF2* and the maternally imprinted inhibitor work in an antagonistic fashion, and a balance between the two is required for normal development. This seems to be more generally the case as well. For normal embryonic development, a balance is required between the activities of maternally and paternally imprinted genes.[21]

Imprinted genes are uniquely vulnerable to molecular mishaps because of their *monoallelic* (one allele) expression. In most genes, which are *biallelic* (expression by both alleles), if something goes wrong with one allele, the other can partially compensate. There can be no such compensation in imprinted alleles. If something goes wrong, it goes wrong in a bigger way than in most genes.[22] The consequences of these epigenetic mishaps are enormous, partly because they occur so early in development, but also because the messed-up imprints are more likely to be transmitted to future generations than other messed-up epigenetic processes. Imprinting has transgenerational effects.

Environmental Effects on Imprinted Genes

There is increasing interest in the effects of environmental toxins on epigenetic processes in general, and recently, genomic imprinting in particular. Here I will focus on one group of toxins called endocrine disruptors. As the name implies, *endocrine disruptors* disrupt physiological processes that involve hormones, typically because they mimic the hormones and bind to their receptors. Some of the most pernicious endocrine disruptors mimic the female hormone estrogen. These include polychlorinated biphenyls (PCBs) and bisphenol A, which is used in the production of plastics (such as the ubiquitous water bottles). Other estrogen-like endocrine disruptors include agricultural weed killers such as atrazine, and fungicides such as vinclozolin.

The effects of endocrine disruptors were first noticed in fish and amphibians, and are a major cause of the declines in some local populations.[23] Fish and amphibians are especially susceptible for two reasons: because they live in aquatic habitats where these chemicals become concentrated, and because their sexual development is influenced more by their environment than is the sexual development of humans and other mammals.[24] For example, endocrine disruptors can cause fish to change sex, resulting in all-female populations.[25] They can also have a dramatically feminizing effect on amphibians resulting in male sterility.[26]

While less dramatic than in fish and amphibians, endocrine disruptors have been linked to a variety of ailments in humans and other mammals. The effects of endocrine disruptors on imprinted genes in mammals are especially well studied.[27] Male mammals, including human males, seem particularly sensitive to developmental errors caused by the effect of endocrine disruptors on imprinted genes, as evidenced by increased rates of prostate cancer, kidney disease, and abnormal testes.[28] In many cases, these problems don't become manifest until adulthood, as in the adult-onset diseases such as the metabolic syndrome. As if this were not alarming enough, it has been recently demonstrated in rats that these defects can be transmitted to future generations.

Male rats exposed to the fungicide vinclozolin in utero have defective sperm and reduced fertility as adults. Their male offspring—with no vinclozolin exposure—also have defective sperm and low fertility, as do the males of the third and fourth generations.[29] Vinclozolin exerts these transgenerational effects by altering the imprinting process during sperm development. The fungicide not only alters normal imprints but establishes new ones in parts of the genome that are not usually imprinted.[30] These new imprints are transmitted through the male line for at least four generations. The new imprints not only affect fertility; they are also associated with a number of adult-onset diseases of the testis, prostate, kidney, and immune system.[31]

These experiments have not been replicated in humans, nor will that ever happen—what prospective mother is going to volunteer for vinclozolin exposure? But they provide compelling evidence that endocrine disruptors are not just a problem for fish and frogs.

The Hybrid Problem

We began this chapter with the mule-hinny puzzle, to which we now return. We should first note that members of the horse family have a remarkable ability to produce healthy young through hybridization. This is true not only for horses and asses, but for zebras too. You can cross a zebra with a horse and get a zorse (male zebra × female horse) or a horbra (male horse × female zebra). That is not at all the norm for mammals. Except among the most closely related species, mammalian hybrids exhibit all kinds of developmental defects and health problems, a phenomenon known as hybrid dysgenesis. Members of the horse family are not immune to hybrid dysgenesis, as is evident in the sterility of the mules and other hybrids. Traditionally hybrid dysgenesis was attributed to genetic incompatibilities. Once two species have genetically diverged sufficiently, any hybrid will suffer because when the two parental genomes are combined in the fertilized egg, they cannot be properly coordinated.

There is undoubtedly much truth to this view, but recent research suggests that it is only part of the story. Hybrid mammals also experience disruptions in the imprinting process, including a complete loss of imprinting for some genes. The loss of imprinting has been especially well demonstrated in rodents such as members of the genus *Mus,* which includes the common house mouse. Alleles that are normally expressed only when inherited from the mother—or from the father, as the case may be—are now expressed no matter from which parent they were derived. This can create a host of problems, beginning very early in development.[32] The problem here is not so much genetic divergence as epigenetic divergence, which leads to problems in epigenetic reprogramming.

Horses and asses have also epigenetically diverged, though not to the point of causing fundamental disruption of epigenetic reprogramming and consequent loss of imprints. Stallions and jackasses bequeath slightly different imprints to their progeny, as do mares and jennies. Hence the crosses, though genetically symmetrical, are not epigenetically symmetrical. The differences between mules and hinnies quite effectively illustrate the power of this epigenetic asymmetry.

Mule Variations

Mules (and hinnies) were first created over three thousand years ago by some enterprising—if somewhat perverse—inhabitants of Mesopotamia, the first recorded example of a parent-of-origin effect. Over the years, other parent-of-origin effects were uncovered, not only in hybrids but in the transmission of numerous sorts of developmental defects, such as Prader-Willi and Turner syndromes. Yet these parent-of-origin effects remained a puzzle, even long after the advent of modern genetics. Mendel's framework and its subsequent elaborations provided insufficient resources for understanding this phenomenon.

It is only very recently, with the advent of epigenetics, that we have at hand an explanation for mules and hinnies and other parent-of-origin effects, now known as genomic imprinting. Imprinting resembles, in some ways, the kind of epigenetic inheritance we explored in Chapter 7. An important difference is that the epigenetic mark in imprinting is not directly transmitted to the next generation as it is in the mouse agouti allele or the *fwa* allele in *Arabidopsis*. Instead, it is erased during epigenetic reprogramming, then reestablished anew. For this reason, imprinting is not considered true epigenetic inheritance, even though imprints are certainly epigenetic and they are inherited, albeit in a different manner than genes and epigenetic marks such as *fwa*. Whether we want to call it epigenetic inheritance or simply another kind of transgenerational epigenetic effect, imprinting clearly calls for an expansion of our notion of biological inheritance. Imprinting is a form of biological inheritance; it just follows different rules than genetic inheritance.

But genomic imprinting is first and foremost a novel form of epigenetic control of the process of development, the process whereby a fertilized egg, or *zygote*, became you or me. We now turn to the more common means by which this process is epigenetically regulated. For most biologists, it is in understanding development, especially very early development, that the biggest payoff for epigenetics will be.

Chapter 10

Sea Urchins Are Not Just to Eat

THE SEA URCHIN, REVILED BY OCEAN REVELERS FOR ITS SPINES but beloved by sushi aficionados for its gonads, has a starring role in developmental biology.[1] Much of what we know about the earliest stages of development comes from studies on sea urchins. Fertilization—the union of sperm nucleus with egg nucleus—was first observed in sea urchins. Sea urchins have also figured prominently in studies of what happens next: a series of cell divisions through which the fertilized egg, or zygote, is transformed into a ball of generic cells called a *blastula*. The cells in the blastula are generic in the sense that they have none of the distinguishing features of any of the cell types found in adult sea urchins, such as blood cells and neurons (sea urchins have neurons but no brains). These blastula cells are also generic in another sense: they generate all of the adult cell types. We call these generic blastula cells *embryonic stem cells*. This is as true for humans as for sea urchins.

The process whereby the zygote gives rise to the blastula, and the blastula gives rise to an animal with spines and gonads—or

brains and gonads, as the case may be—is among the most won-
drous in the universe. This process is also one of the most difficult
for the human mind to digest. Our intuition, so useful in other scientific contexts, tends to fail us here, tends to lead us astray. There
is also a lot at stake here. The ultimate goal, after all, is to understand how we come to be what we are, how you came to be you. It
is not surprising, therefore, that this process, which we will simply
call *development*, is a subject with a long history of dispute.

There are a number of points of contention, but for our purposes we can divide the disputants into two camps. In the first,
we have those who assert that, despite appearances to the con-
trary, the zygote actually contained you. This is known as preformationism, which in its most extreme form asserts that your
development was merely a matter of growth. The extreme form of
preformationism is also the most primitive; far more sophisticated
versions were developed during the eighteenth and nineteenth
centuries.[2] The more sophisticated versions of preformationism
assert that you are latent in the zygote and your development is
the process of the "latent you" becoming the "manifest you." This
latent you need not resemble the manifest you at all. But—and
this is the essence of preformationism—the particular form that
constitutes you is entirely present, however latently, in the zygote.
Nothing in the environment contributes to your adult form. Nor
did your manifest you come to exist as a result of development; it
was there from the outset.

The last claim in particular distinguishes preformationism from
the second type of developmental explanation, known as epigenesis.[3] From the epigenesist perspective, you do not exist prior to
development, either manifestly or latently. Rather, development is
the process whereby you come to exist. Development is not just a
matter of unfolding; it is a creative process. This is not to deny that

the genes and other biochemicals in that zygote are essential for your development—they most certainly are. But they don't contribute to your development by being a preformed you.

In the earliest (that is, seventeenth- and eighteenth-century) versions of preformationism, the preformed you was usually thought to reside in the unfertilized egg. Moreover, each egg was considered to be something like a set of Russian dolls, containing within itself the forms of all subsequent generations in increasingly diminutive sizes. Eve's eggs contained within them all subsequent humans, including you. Though this seems absurd to us now, it was not, given the technology and knowledge existing at that time, a ridiculous hypothesis. Moreover, one of its considerable virtues was its ability to explain development without invoking any supernatural principles that violate the basic scientific principle of naturalism (as opposed to supernaturalism). Preformationists did not need any spooky stuff to explain how your mother's egg became you. There is nothing mysterious about growth.

Advocates for epigenesis, however, claimed that despite its naturalism, preformationism was simplistic at best, and at worst downright wrong-headed. Epigenesists were generally more attentive than the preformationists to the complexities of development, but they had problems of their own. Most significantly, they could not account for the orderliness of this complex process—or how this ordered complexity could result from more simple, seemingly homogeneous conditions—without recourse to spooky stuff. For the preformationists, complexity was unproblematic; it existed from the outset, from the dawn of creation. Early epigenesists, however, like Darwin in a different context, had to explain how you get something very complex from something simple like a zygote. They could not find the explanatory principles they needed among the known physical laws, most notably Newton's physics, so they posited an additional something that living things have and non-living things lack. The conceptions of this additional something varied greatly, but they had in common the notion that it is not material, and that it could only be demonstrated by pointing to the process itself. This does not count as a scientific explanation.

By the last few decades of the nineteenth century, experimental methods had improved enough to actually test some of the tenets of both camps. Two German scientists figured prominently in the debate at this point, Wilhelm Roux and Hans Driesch.

The Experiment and Its Aftermath

Driesch chose sea urchins for his study animal for the same reason that others before him had chosen sea urchins to study fertilization: their large, largely yokeless eggs. Sea urchin eggs are much larger than those of most other animals, including frogs and humans. This is especially important for monitoring not only fertilization but also the first few cell divisions, which occur within the confines of the zygote. As such, with each cell division, the cells become progressively smaller. But even after the first few cell divisions, the cells in the sea urchin embryo are relatively large, and hence easy to see with the microscopes available at that time, compared with cells in frog or human embryos at the same stage. The fact that sea urchin zygotes have very little yolk means that these cells are nearly transparent, which is also a plus.

A central issue in the debate at this time was, how do you get all the different cell types—blood cells, skin cells, cone cells, and so on—from an egg cell that doesn't look anything like any of them? The state-of-the-art preformationist account in the 1890s, the one advocated by Roux, proposed that the chromosomes in the fertilized egg or zygote contain all of the determinants of the adult form.[4] Parts of the chromosomes are then increasingly parceled out during each subsequent cell division, until all of the cell types have become fully differentiated. Whether the differentiated cell becomes a neuron, a muscle fiber, a blood cell, or whatever type depends on which particular bits of chromosome it contains. The idea appealed to many because it was straightforwardly mechanical.

Both Roux and Driesch set out to test this idea, but only Driesch succeeded in mastering the technical details.[5] Driesch experimentally intervened right after fertilization, during those first few cell divisions. He managed to separate the cells in sea urchin embryos when they were only at the two- to eight-cell stages. To his surprise, each of the separated cells from the two-cell stage developed into a complete sea urchin larva, as did some from the four- and eight-cell embryos. According to Roux's preformationist account, any cell isolated at the two-cell stage should have half of the chromosomal complement and could thus only become half of a sea urchin. As a result, Driesch's sea urchin experiments amounted to a decisive refutation of Roux's preformationism, and Driesch became an advocate for epigenesis.

Driesch reasonably concluded that in the very early embryo, each cell could somehow regulate its development so as to become a whole embryo. Processes such as Driesch envisioned for development, we now call self-organizing.[6] Driesch then carried out another set of experiments, at much later stages of development, which were of even more significance. He ingeniously manipulated the position of cells such that those that would normally have become, say, spines, were moved to a position appropriate for mouthparts. According to Roux's preformationism, the resulting embryo should have been a mess, with spines sticking out of its mouth. Instead, Driesch obtained normal embryos.

Driesch concluded that the potential of each cell is not determined by what chromosomes (genes) are in its nucleus, but rather on its position in the embryo. More generally, Driesch began to develop an epigenetic framework in which he anticipated the modern concept of gene regulation by the cellular environment. He proposed that cellular stuff outside of the nucleus (where the genes reside) influences what the nucleus (genetic material) does, which in turn influences what the cellular stuff does, and so on, in a reciprocal manner.[7] This notion of reciprocal causation, or feedback—in which an action is both cause and effect—was new to science. It is now a fundamental principle in biology.

Like Roux, Driesch had commenced his experiments with the thoroughly naturalistic vision of explaining development using only the laws of physics and mathematics. As such, he found his results somewhat disillusioning. In fact, he became so overwhelmed by the complexity of development that he abandoned not only his simpleminded naturalism, but naturalism as a whole. He felt compelled to invoke a soul-like principle, which he called "entelechy" following Aristotle, to explain this complex process.[8] He eventually abandoned biology for philosophy, to the detriment of both.

The Death and Resurrection of Preformationism

Driesch's experiments doomed the existing incarnation of preformationism, but not preformationist thinking. In part, the reincarnation of preformationism reflected the inability of advocates for epigenesis to come up with a credible nonspooky mechanism. But there was another force behind the resurrection of preformationism: its intuitive appeal. The human mind seems predisposed toward preformationism; it taps into some naïve preferred modes of thought for scientists and nonscientists alike, when it comes to explaining complex phenomena like development (or evolution).

There are two related intuitions that undergird preformationism (as well as creationism). The first intuition concerns the "pre" in "preformationism": the complexity must be there from the get-go, in the egg (or the mind of God). Call this the "complexity intuition." The second intuition concerns the "form" in "preformationism": the preexisting form in the egg (or the mind of God) directs the development of the adult form. Call this the "directorial intuition."

A naturalized form of epigenesis must combat both intuitions without recourse to spooky stuff. Against the intuition that only complexity begets complexity, epigenesists must successfully argue that complexity can result from relatively simple initial conditions. Against the intuition that complexly ordered processes such as development require a central form-giving director, epigenesists must demonstrate that this order can result from local interactions at the cellular level, given certain initial conditions.[9] This is not an easy task. Even Driesch, while anticipating the essential concepts of feedback, reciprocal causation, and self-organization, eventually copped out by resorting to the spooky notion of entelechy. Hence, many biologists, while acknowledging that Driesch's experiments won the battle for epigenesis, could not bring themselves to concede that he had won the war. They felt vindicated in this regard by the rise of modern genetics.

Originally the science of genetics—as practiced by, for example, Morgan—developed completely independently of developmental biology. It was inevitable, however, that geneticists would eventually tackle development. They did so with a strong preformationist bent, largely unaware of the previous work by Driesch and others in developmental biology. On the new, genetically inspired preformationist account, your genes contain the complex form that is you, through which they direct the process of your development. Genetic preformationism was successfully packaged through a series of intuitively appealing metaphors. First came the metaphor of the "genetic blueprint," then that of the "genetic recipe," and finally the "genetic program." Some combination of the recipe and program metaphors remains popular; they have in common the notion that the genes provide instructions, which cells execute. In genetic preformationism, the executive gene is scaled up to become the executive genome.[10]

These recipe/program metaphors are attractive because they connect the basic intuitions common to all versions of preformationism to human artifacts with which we are all familiar, from cakes to graduation ceremonies.[11] Whatever their intuitive appeal, these metaphors cannot withstand even the most cursory scrutiny. You couldn't cook up a single cell, much less a human being, given the instructions in the genetic recipe. Much of what you need to know lies elsewhere. More to the epigenesist point, most of the information in the recipe that goes into making you is not there from the outset. Rather, development is the process whereby this information comes to exist.[12] The recipe is written during development, not prior to development.

The same goes for the metaphor of a genetic program. But the notion of a genetic program also suffers from another fatal flaw: the software-hardware distinction. The genes are supposed to comprise the software and the rest of the cellular constituents the hardware whose operation the genes instruct. But as we have seen throughout this book, our genes are as much part of our hardware as any other biochemicals, and as much instructed as instructors. In fact, the science of epigenetics makes sense only when genes are viewed as biochemical hardware.

From Stem Cells to Cone Cells

One of the most vexing problems for all versions of preformation-ism—from the most primitive (Eve's eggs) to the most sophis-ticated (genetic programs)—is explaining how a ball of identical generic cells can give rise to specialized cells such as cone cells, neu-rons, muscle cells, and so forth. This process is called *cellular dif-ferentiation*. You will recall that Roux, the preformationist, thought differentiation was the result of the progressive partitioning of genes during development. The zygote has all of the genes, while cone cells retain only a small subset and muscle cells a different small subset. But we now know that all of the cells in our body are genetically identical.[13] Cone cells have the same genes as liver cells, muscle cells, or any other cell type. What makes cone cells differ-ent from, say, heart muscle cells, is differences in gene expression, not differences in genes. These differences in gene expression are caused by epigenetic processes.

Think of the zygote in terms of a particular commodity: poten-tial. Zygotes have lots of this potential; in fact, they have as much potential as it is possible to obtain. What sort of potential? The potential to become—through numerous intermediary states—any one of the more than two hundred cell types in the human body, plus the placenta. They are said to be *totipotent*, which basi-cally translates as "total potential." By the blastula stage (about 128 cells), the component cells have lost the capacity to become placental cells, but they can become any of the other two hun-dred plus cell types. These are the *embryonic stem cells*, which are said to be *pluripotent*. From the blastula stage on, some of that potential to become other cell types gets lost with each suc-cessive cell division. The daughter cells lose some of that poten-tial in the mother cell; they become more limited in what they

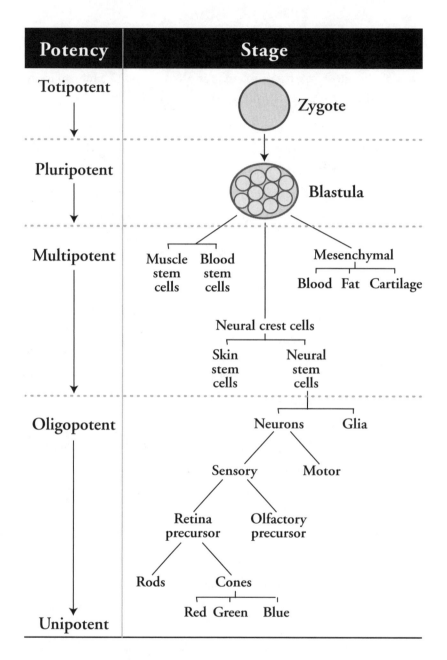

Potency	Stage

Totipotent — Zygote

Pluripotent — Blastula

Multipotent —
Muscle stem cells   Blood stem cells   Mesenchymal
Blood Fat Cartilage

Neural crest cells
Skin stem cells   Neural stem cells

Oligopotent —
Neurons   Glia
Sensory   Motor
Retina precursor   Olfactory precursor
Rods   Cones
Red Green Blue

Unipotent

A schematic diagram of cellular differentiation from zygote to terminal differentiation of cone cells. Oligopotent cells have less potential than multipotent cells but more than unipotent cells. The branching within the oligopotent region illustrated here is highly speculative. Diagram by the author.

can—through numerous intermediary states—become. When that potential gets narrowed to a certain point, where the mother cells give rise to daughter cells that can only become a certain class of cell types—say, blood cells versus neural cells—we call them *somatic stem cells*. (Somatic stem cells are often misleadingly referred to as "adult stem cells" in the media.) Somatic stem cells have lost much of that potential possessed by the embryonic stem cells, but they still retain a large amount of potential, compared with cone cells or heart muscle cells. They are said to be *multipotent*, which means, roughly, the capacity to become multiple cell types—in contrast to both no potential (for example, cone cells) and almost total potential (as in embryonic stem cells). There are a number of types of somatic stem cells, including neural stem cells and blood stem cells.

The transition from pluripotent embryonic stem cells to multipotent somatic stem cells, such as neural stem cells, is an epigenetic process during which an increasing number of genes are permanently inactivated (while other genes are newly activated). Differentiation continues past the multipotent somatic stem cell state—through further progressive epigenetic inactivation—until terminal differentiation into one of the two hundred plus cell types, such as cone cells or heart muscle cells. Cone cells and heart muscle cells are the end of the line differentiation-wise: cone cells can only beget other cone cells; heart muscle cells can only beget heart muscle cells. They have no more potential to become anything else.

A Third Form of Epigenetic Regulation

Both DNA methylation and histone modifications figure prominently in cellular differentiation, but there is another epigenetic mechanism at work here as well, which involves RNA.

We have already met one type of RNA involved in epigenetic regulation called Xist, which plays a central role in X-chromosome inactivation. Most of the RNAs with epigenetic functions are much smaller than Xist, however. Many, in fact, are extremely small. One important group of these tiny noncoding RNAs is called *microRNAs*.[14]

Epigenetic gene regulation by microRNAs works quite differently than either methylation or histone binding. Most significantly, microRNAs act later in the protein synthesis process. Recall from Chapter 2 that protein synthesis occurs in two stages. During the first stage, called transcription, messenger RNA (mRNA) is constructed from the DNA template. During the second stage, called translation, a protoprotein is constructed from the RNA template. Most epigenetic gene regulation occurs at the first stage, usually by inhibiting transcription. MicroRNAs, in contrast, exert their influence during the second stage, translation.

Though transcription is highly regulated, it is often the case that there are too many mRNA transcripts from a particular gene for a cell's purposes. If a cell "decides" that this is the case, it deploys microRNAs to remedy the situation. The microRNAs identify the particular mRNAs that are overabundant and mark them for destruction. The microRNAs must physically bind to the messenger RNAs at a site where they are complementary. The microRNA doesn't have to be perfectly complementary, just complementary enough to stick to the much larger mRNA. That means there can be many fewer types of microRNAs than there are types of mRNA transcripts. Sometimes the binding of a microRNA to an mRNA is sufficient to block the translation of the mRNA into protein. Sometimes the microRNA, in addition, attracts proteins, including enzymes, which actively degrade the mRNA. In either case, the net result is that there is less mRNA available to serve as a template for protein construction. This microRNA-based form of gene regulation, known as *RNA interference*, functions as a way to fine-tune the amount of protein that is made from a particular gene.[15] Some have likened it to a light dimmer.

MicroRNAs play a major role in normal cellular differentiation.[16] It appears that one of the primary functions of microRNAs at the cellular level is to stabilize the differentiation state of the cell. One study suggests that microRNAs don't drive the differentiation but rather prevent dedifferentiation.[17]

Differentiation Is Reversible

You may have noticed that the process of differentiation superficially resembles Roux's preformationist schema. Rather than a progressive partitioning of genes, there is a progressive partitioning of gene expression. But the resemblance is only superficial and the differences deep. The progressive partitioning of gene expression by epigenetic inactivation is not at all preformationist. Rather, as Driesch inferred, the fate of each cell, even well beyond the pluripotent stage, is largely determined by its position in the embryo and the nature of its neighbor cells with which it (chemically) interacts. These intercellular interactions influence the environment within the cell, which in turn influences which genes are epigenetically activated or inactivated. Hence, if you move cells around, especially during early stages of embryonic development, their fates change, just as in Driesch's sea urchins, where cells normally destined for spinehood became part of the mouth when moved there.

The fact that differentiation is reversible also counts against any preformationist account. Under certain conditions, cells can *dedifferentiate*, that is, move back up toward "stemness." This occurs naturally in injury repair, most spectacularly in amphibians, which can regenerate whole limbs and other organs.[18] Dedifferentiation of skin, muscle, and bone cells is an essential part of this process. In mammals, dedifferentiation occurs in response to injury to cartilage and the peripheral nervous system.[19] Some researchers seek to extend the range of dedifferentiation-based injury repair in humans based on amphibian models.[20] But the most exciting potential for dedifferentiation therapies comes from stem cell research.

Stem cell biologists can now take a cell from the skin of an adult and, through biochemical subterfuge, transform it into the equivalent of an embryonic skin cell.[21] Moreover, they can now take this skin cell–derived embryonic stem cell and cause it to redifferentiate into a neuron.[22] Dedifferentiation is also one path to cancer, as we will see in the next chapter. But cancer cells themselves can be used to generate embryonic stem cells, in what is perhaps the most dramatic example of epigenesis.

Researchers were able to transplant human malignant melanoma cells into a chick embryo, a seemingly perverse procedure destined for failure.[23] You would expect the melanoma cells, if they become integrated with the embryonic cells, to cause lots of problems. But in fact, the chick embryos were not compromised by the cancer cells. It's not that the cancer cells died; they survived and divided at the normal rate. But the melanoma cells didn't form tumors. Rather, the descendents of these cells actually assumed the identity of the normal cells from which the tumor cells were ultimately derived, in this case neural crest cells.[24] They also migrated into the proper position for neural crest cells. It's quite remarkable, really: you can use cancer cells to generate normal cells by exposing them to stem cells.

What happened to the cancer cells? Somehow, through interactions with the embryonic chick cells, the human cancer cells were epigenetically altered. These alterations caused the melanoma cells to dedifferentiate into stem cells, which then were able to redifferentiate into normal cells. This wasn't the unfolding of a genetic script, so much as the writing (and rewriting) of it.

Epigenesis Without the Spooky Stuff

Epigenetics, it seems, provides the long-awaited naturalistic mechanism for epigenesis and should therefore spell the doom for preformationism. But the intuitive appeal of preformationism gives it a whack-a-mole capacity for resurrection. Every time the latest version gets crushed, a new version pops up to replace it. I will call the latest version of preformationism the "genetic-epigenetic program."[25] The genetic-epigenetic program metaphor acknowledges the central role of epigenetic events in development but views them through a preformationist lens. In essence, the idea is that the epigenetic events described earlier are programmed by the executive genome.

All of the problems with the "program" metaphor (as in "genetic programs") apply to the notion of a genetic-epigenetic program—plus an additional one: in what sense are these epigenetic events programmed? Certainly not in the sense of "program" that most readers will bring to the table: a recipe-like set of instructions. As we have seen, the epigenetic changes in gene expression that determine a cell's fate are largely determined by the position of that cell in the developing embryo. Therefore, it would be more apt to say that the genes are programmed by cellular interactions.

There are other, very minimalist senses of "program" that have become popular in the fields of artificial intelligence and artificial life.[26] On this minimalist notion of program, a program provides a few basic rules, and the robots or cellular automata take it from there through interactions with their neighbors and the rest of their environment. But the idea of a central director is gone. This minimalist sense of program looks a lot like epigenesis. So the preformationist notion of an epigenetic (or genetic) program is either false or too squishy to be distinguished from epigenesis. In either case, we would be better served to drop the "program" metaphor altogether, and with it, the temptation to think of genes as software. Better to treat genes as concrete, material (biochemical) things, if we hope to understand the epigenetic processes that regulate them during differentiation.

A Note on the Stem Cell Controversy

Because of the protean nature of embryonic stem cells, they have the potential to revolutionize medicine. In principle you can place embryonic stem cells in any damaged part of the body, including the brain, and they will proliferate and differentiate into replacements for the damaged cells whatever they are. Heretofore untreatable conditions, such as injury to the spinal cord, can potentially be remedied in this way. Embryonic stem cells could help those who have been paralyzed for years to walk again—just one of countless potential applications.

Somatic stem cells can be utilized in some of the same ways. For example, neural stem cells might be able to do the job for spinal cord injuries. That would be a great boon because somatic stem cells are much easier to come by than embryonic stem cells. Until very recently, embryonic stem cells could only be harvested from blastula-stage embryos, a practice opposed by many on religious grounds. But we retain reservoirs of somatic stem cells even into adulthood, which is why somatic stem cells are often misguidedly referred to as adult stem cells (a misnomer since somatic stem cells are present in all but the earliest stages of embryonic development). It is therefore much easier to obtain somatic stem cells, and there is little opposition to this practice. But there are many applications for which embryonic stem cells remain more effective than somatic stem cells.[27] This is why the recent studies demonstrating that stem cells can be induced artificially met with such fanfare. Here, potentially, is a way to generate more embryonic stem cells through dedifferentiation, one that won't provoke the ire of opponents of embryonic stem cell research. But it will take years for scientists to convert this proof of principle into a viable technology for generating embryonic stem cells.[28] In the meantime, embryonic stem cells harvested from embryos (blastulas) remain the best hope for many.

Opponents of embryonic stem cell research are motivated by a religious form of preformationism. They believe that a human soul is created when sperm meets egg. Secular versions of preformationism must also concede that the zygote is a human being, with all of the ethical implications that entails. From an epigenesist perspective, though, it makes no sense whatsoever to think of a zygote as a human being. A human being may or may not come to exist as a result of the processes initiated in the zygote. And zygotes produced by humans certainly have a much greater chance of becoming humans than those of, say, sea urchins; but that doesn't make them actual humans. The same goes for the undifferentiated ball of cells that are harvested for embryonic stem cell research.

So at what point does the embryo become a human? Epigenesists can provide no hard and fast answers to that question. They can only note that human development is the process of becoming human, not the process of a preexisting latent human becoming a manifest human. Therefore, humanness is a matter of degree, not an absolute. It is up to us collectively as a society to decide how we should treat varying developmental degrees of humanness.

When Intuitions Lead Us Astray

From the outset, preformationism has had two considerable advantages over epigenesis: its intuitive appeal, and its naturalism. The drawback of preformationism has always been its failure in the experimental realm. Improved microscopes doomed the first (Eve's eggs) version of preformationism. Driesch's sea urchin experiments doomed Roux's second-generation version. Despite these failures, preformationism remained viable because of its intuitive appeal, and because epigenesists like Driesch kept resorting to spooky stuff. When geneticists entered the fray, preformationism got a new boost, as genes seemed to provide an attractive preformationist mechanism, eventually expressed through two seductive metaphors—genetic recipes and genetic programs. Because it is purely metaphoric, this third-generation preformationism is inherently harder to put to the test. But epigenetics, properly construed, should spell the doom of preformationism.

The root of the term "epigenetic" is "epigenesis"—not, as is often supposed, "genetic."[29] Conrad Waddington, who coined the neologism in the 1940s, had in mind a naturalistic (nonspooky) form of epigenesis in which genes would play a prominent role.[30] But the role that Waddington envisioned for genes much more closely resembled the executive cell view than that of the executive gene(ome).[31] For Waddington, genes are as responsive to their cellular environment as their cellular environment is to them.[32]

Much has happened to the science of epigenetics since Waddington, and the temptation to view epigenetics through a preformationist lens remains powerful. The genetic/epigenetic program metaphor in particular is deployed effectively to this end. But modern epigenetics, as evidenced in the process of cellular differentiation, undermines preformationism as genetic program, most fundamentally, by challenging the view that genes are software that instructs the cellular machinery, viewed as hardware. Modern epigenetics makes sense only if genes are viewed as hardware, like other cellular constituents—as much instructed as instructor, as much directed as director, as much effect as cause.

Development does have a programmatic feel to it, but development is not programmed, in any but the squishiest sense of the term. Differentiation, for example, is an orderly process resulting from local cell-cell interactions. Whether a particular embryonic stem cell gives rise to cone cells or heart muscle cells depends on the history of these local interactions. Moreover, stem cells can induce dedifferentiation of differentiated cell types, not least in cancer cells. This dedifferentiation is often referred to as reprogramming.[33] But here it is especially clear that the "reprogramming" isn't done *by* the genes but *to* the genes. The same is true of the so-called programming of normal cellular differentiation. But all this talk of programming and reprogramming is a diversion from the concrete cellular interactions that are at the root of not only normal development but also cancer, which is the subject of the next chapter.

Chapter 11

Pray for the Devil

THE ENGLISH PHILOSOPHER THOMAS HOBBES IS PROBABLY BEST
known for his pessimistic assessment of human existence as "nasty,
brutish and short."[1] I can only imagine what he would have said
about the lives of Tasmanian devils, compared with which the
human condition is idyllic. For "T-devils," those feisty creatures
from Down Under's down under, life is certainly nastier, more
brutish, and shorter. And recently, it's gotten a lot shorter.

A Tasmanian devil is born in a highly undeveloped state, even
by marsupial standards, which are much lower than those of most
mammals. The newborn T-devil is tiny, the size of a grain of rice.
So the trip from the vagina to the pouch, though only 3.5 inches, is
an epic journey through a forest of giant hairs. It is also a life-and-
death race. A mother T-devil gives birth to thirty to forty young,
yet she has only four nipples. The first four babies to the nipples
win this first challenge; all of the others die. Not surprisingly, once
they arrive at the nipple, they clamp on like a tick and don't release
their grip for weeks.

For those who make it this far, the ensuing few months are relative bliss. After they let go of the nipple, there is the security of the pouch, and later the lesser security of a burrow. But then the young T-devils must strike out on their own in their quest for food and a mate, endeavors fraught with turmoil and conflict.

The first part of the scientific name for the T-devil is *Sarcophillus*, Latin for "flesh lover," and they do consume a wide variety of fleshy foods, most of it carrion. At a large carcass, such as that of a kangaroo, many T-devils may congregate. The ensuing melee is a noisy affair; one of the sounds produced by agitated T-devils is their shudder-inducing scream, believed by many to be the source of their devilish reputation. (Others attribute the devil moniker to the fact that when they become emotional, their ears get bright red.) T-devils can also produce foul odors during these aggressive interactions, said to rival those of a skunk.

But T-devil aggression does not end with their vocal and olfactory repertoire; they also deliver nasty bites with a force which, pound for pound, is the greatest among mammals, exceeding even that of the spotted hyena.[2] T-devils often have scars on their rump because they wisely back into the carcass to protect their head and face. The backing-in technique works only somewhat, however; eventually, T-devils do get bit in the face during these encounters, when hunger wins out. Moreover, bites to the face occur during courtship, which is also a violent affair. These are essentially solitary creatures who are not real comfortable in each other's company, whether seeking food or sex. Fortunately, T-devils have an amazing capacity to heal quickly, even huge gaping wounds—until recently.

The Cancer from Hell

In 1996, a wildlife photographer, working in Mount Williams National Park, observed a number of T-devils with strange growths on their face and mouths. Soon after, wildlife biologists in Tasmania began to see more and more afflicted individuals. By 2002, this affliction was an epidemic throughout most of the T-devil range.[3] The growths, it was discovered, were a strange kind of cancer of the mouth and face, which was labeled devil facial tumor disease, or DFTD. This rapidly growing cancer eventually obstructs the mouth to a degree that the affected individuals die of starvation, usually within several months. The cancer has caused T-devil numbers to drop precipitously in the last decade, and there is no sign it is abating. If the trend continues, the T-devil will soon be extinct.

Why has this cancer so suddenly afflicted so many T-devils? Cancers are not like viruses—we don't, fortunately, see cancer epidemics—yet that is precisely what appears to be happening to T-devils. A cancer epidemic can occur only if the cancer is infectious. But this infectious cancer is not transmitted through a virus or any other vector; it is directly transmitted from one T-devil to another during those aggressive interactions at the carcass or during what passes for T-devil courtship. When an infected individual bites another T-devil, some of the cancer cells are transferred from the biter to the bitten.[4] This is truly nightmarish stuff. DTFD is a parasitic cancer.

There is a sense, though, in which all cancers are parasitic. Our immune systems sometimes treat cancer cells like external invaders. Only those cancer cells that through various ruses evade or disable the immune system eventually form tumors. Those cancer cells that evade the immune response compete with the normal cells that surround them for the body's resources, as does any parasite.

But the DFTD cancer should be at a huge disadvantage relative to cancers that originate within a T-devil's own body: cancers that originate outside of the body should be much easier for the T-devil immune system to identify, and hence destroy. Yet the T-devil immune system does not appear to mount any defense whatsoever against these foreign invaders. Therefore, there must be something wrong with the T-devil immune response. The problem is not a generally deficient immune response. The T-devil immune system responds robustly to most challenges, as you would expect: given their diet and frequent wounds, a T-devil with a generally understated immune response would not be long for this world.

The problem, it seems, is in the recognition phase. The T-devil immune system just isn't able to identify these foreign cancer cells as foreign, as *nonself*. The self-nonself distinction is fundamental to the immune response. It is the reason we reject donated organs, even those from close relatives. Massive doses of immunosuppressive drugs are required for organ transplants, including skin grafts. Sometimes, when the recognition process breaks down, the immune system attacks healthy cells that have the proper *self* markers. The result of this hypervigilance is autoimmune diseases such as rheumatoid arthritis or lupus. The T-devil has the converse problem of immune misidentification; its immune system is too permissive.[5]

This blind spot in the immune response is thought to stem from a genetic bottleneck that occurred sometime after the last ice age, perhaps as recently as the twentieth century. At some point, the T-devil population may have been reduced to just a few individuals, which inbred to a degree that eliminated most of the genetic variation in T-devils even many generations removed. Something similar occurred in cheetahs, which also evidence little genetic variation, accept skin grafts from other cheetahs, and presumably would be vulnerable to a contagious cancer of this sort.[6]

DFTD most closely resembles a cancer in dogs called canine transmissible venereal tumor (CTVT), which is also directly transferred from one individual to another, in this case through sex.[7] Here, too, the immune system fails to recognize the foreign tumor cells as foreign. But dogs with CTVT eventually do mount an immune response that completely eliminates this cancer.[8] (Once an affected dog has recovered, it has lifelong immunity to further infections by this cancer.) Unfortunately, T-devils are not so lucky, which indicates that there is more wrong in the T-devil immune response to DFTD than a defective recognition phase.

Cancer and Stemness

As bizarre and unusual as the T-devil cancer is, the cancer cells themselves are fairly typical cancer cells. For example, they are poorly differentiated; that is, they bear certain resemblances to somatic stem cells. As is also characteristic of cancer cells, however, DFTD cells have some of the attributes of the type of cell that they should have become. Also typical is the rearrangement of the chromosomes in these cancer cells, which, in addition, have completely lost one pair of chromosomes.[9] Chromosomal deletions (and additions) are common in cancer cells, no matter the source.

There are two main views as to the cellular transformations that constitute cancer. The traditional view, which was briefly discussed in Chapter 10, is that cancer cells are derived from fully differentiated cells, such as neurons or skin cells. As a result of dedifferentiation, these cells have regained a stem cell–like capacity to proliferate.[10] This dedifferentiation would also explain why these cancer cells retain some of the characteristics of their source cells. The T-devil cancer cells are thought to have originated from a particular kind of neural tissue that controls the endocrine (hormonal) system, based on certain chemical signatures of this tissue.[11]

Recently, an alternative to the dedifferentiation account has been proposed. According to this view, cancer cells are actually derived from somatic stem cells gone bad.[12] On this stem cell theory of cancer, the reason cancer cells resemble stem cells is that their mother cells *were* stem cells. After their birth, they took a wrong turn, converting from normal somatic stem cells to cancer stem cells. Actually, only a minority of the cancer cells retain stem cell properties. Like normal stem cells, these cancer stem cells undergo asymmetric cell division, resulting in one cancer stem cell and one more differentiated cancer cell. The more-differentiated cancer cells then undergo the symmetrical form of cell division typical of all non–stem cells. The net result is a tumor that consists of a small number of cancer stem cells, and a large number of cancer cells that are—to varying degrees—more differentiated. From this perspective, the goal of any therapeutic intervention should be to knock out the relatively few cancer stem cells.

I can summarize the differences between the dedifferentiation and stem cell theories of cancer in this way: on the dedifferentiation account, cancer cells move backward toward stemness; on the stem cell account, cancer cells move forward from stemness. The two theories of cancer are not mutually exclusive. Many prostate cancers show signs of dedifferentiation.[13] On the other hand, cancers of the blood, such as leukemia, may be better explained by the stem cell theory of cancer.[14]

Cancer Genes and Wayward Chromosomes

The dedifferentiation and stem cell theories concern what I will call "cancer dynamics." The hypotheses under consideration from here on concern the mechanisms underlying these dynamics. Most of these hypotheses are compatible with either the dedifferentiation theory or the stem cell theory of cancer dynamics.

What, initially, causes the cancer cells to become cancerous? For the last forty-plus years, the answer to this question has been some sort of genetic alteration in a single cell, which causes it to proliferate abnormally. Further mutations accumulate in the expanding cell population, leading to genetic heterogeneity in the cancer. These different genetic clones compete with each other through further proliferation, becoming increasingly virulent, culminating in metastasis. So cancer is, first and last, from its initiation to metastasis, a matter of genetic alterations. This is known as the *somatic mutation theory* (SMT) of cancer.[15] According to SMT, cancer is a case of evolution on a small scale.

Since the advent of SMT, more than a hundred human oncogenes (*onco* = "cancer") have been discovered; when mutated such that they are expressed at abnormally high levels, oncogenes promote cancer-like cell proliferation. Moreover, more than thirty tumor suppressor genes have been discovered; as their name implies, these suppress cell proliferation. Mutations in tumor suppressor genes that cause them to be less suppressive are also associated with cancer. The mutations in these genes may be spontaneous—that is, essentially random—or they could occur in response to environmental toxins such as tobacco smoke, pesticides, or ultraviolet radiation, which we refer to as carcinogens.

From the perspective of SMT, a carcinogen is a mutation inducer.

Cancer therapy should be directed toward eliminating the mutated cells. If the source of the mutated cells is cancer stem cells, they should be the focus of treatment. The standard arsenal of cancer treatments, including surgical removal, radiation, and most forms of chemotherapy, are based on the SMT model.

Colorectal cancer is the poster case for SMT.[16] This cancer is initiated by a mutation in an oncogene, and each stage in its progression is accompanied by further mutations. The T-devil cancer seems to fit nicely within SMT as well. DFTD is the winner of the clonal selection process, which in addition has evolved a means to be transmitted from one individual to another. But the transmissibility of the T-devil cancer is actually not predicted by SMT. The process of transmissibility involves adaptations to the immune response, whereas SMT is primarily focused on oncogenes and tumor suppressor genes, none of which are involved in such adaptation. Moreover, immune-based forms of therapy were, rather, motivated by a perspective on cancer that differs substantially from SMT, which I will discuss later.

A second genetic theory of cancer is even older than SMT but never as popular. This theory places primary emphasis on the chromosomal abnormalities so characteristic of cancer cells, among them the loss or gain of entire chromosomes. The altered chromosome number is called *aneuploidy*, so this is often called the "aneuploid theory of cancer."[17] According to SMT, aneuploidy is a secondary effect of cancer. Advocates of the aneuploid theory, however, see the chromosomal rearrangements as primary. The aneuploidy hypothesis proposes that the initiation and progression of cancer is due more to the abnormal chromosomes than to mutations in particular oncogenes.

Aneuploidy messes up the regulation of many genes, which leads to further aneuploidy, which leads to even more messed up gene regulation, and so on. One of the deviant traits that results from this disregulation is an increase in the proliferation of the affected cells. But what gets the process going? According to the aneuploid hypothesis, it's a problem with genes involved in maintaining the integrity of the chromosomes during cell division.[18] The progression of a cancer, on this view, is due to the progressive disruption in gene regulation that results from increasing aneuploidy. In support for this argument, advocates of the aneuploid theory cite the fact that cancer cells do not mutate at higher rates than normal cells but cancer cells show a much higher rate of chromosomal rearrangements than normal cells.[19]

Like SMT, the aneuploid hypothesis is neutral as to whether the initial chromosomal destabilization occurs in somatic stem cells or in fully differentiated cells. Nor does this theory offer different therapeutic options.

The T-devil cancer presents a problem for this hypothesis, though not for lack of aneuploidy: T-devil cancer cells have plenty of aneuploidy. The problem is that DFTD cells are all aneuploid in the same way. Also problematic for the aneuploid hypothesis is that they have been aneuploid in the same way for many years. DFTD is extremely stable at the cellular level. Indeed, DFTD is a cell lineage that is much older than any living T-devil.[20] This shouldn't happen on the aneuploid account. The positive feedback process described earlier—between increasing aneuploidy and increasing gene disregulation—cannot be arrested; quite the opposite, it can only accelerate. As such, the aneuploid theory predicts ever-increasing chromosomal rearrangements and increasing variability in the chromosomal arrangements in the cancer cells of a particular tumor.

The lack of variation in DFTD cells is also a problem for SMT. But then, this lack of variation in DFTD cells, along with their transmissibility, is probably what most distinguishes T-devil cancer from more typical cancers of internal origin. Perhaps the two qualities—cellular stability and transmissibility—are related. It is interesting to note, in this regard, that the cells of canine transmissible venereal tumor (CTVT) have been stable for hundreds, perhaps thousands, of years. In fact, CTVT may be the oldest mammalian cell lineage.[21]

The Epigenetic Dimension

The somatic mutation theory and the aneuploid theory of cancer both focus primarily on genetic alterations. Both were also formulated before the advent of epigenetics. Once cancer researchers began looking for epigenetic alterations in cancer cells, they found them. First, it was noticed that the genes of cancer cells have characteristic changes in their methylation patterns, including an overall reduction in methylation.[22] This global hypomethylation is one of the best early predictors of cancer. Genes that are normally repressed become active, including oncogenes. Subsequently, there are also specific changes in the methylation of oncogenes and tumor suppressor genes. Other epigenetic alterations are common in cancer as well, including the unbinding of histones to DNA, which causes an increase in the activity of the affected genes.

Advocates for SMT and the aneuploid theory of cancer do not dismiss the role of epigenetic processes in cancer but consider them secondary to the genetic alterations. Other researchers, though, see the epigenetic alterations as primary in many cases.[23] On the epigenetic view, cancer is most fundamentally the result of defective gene regulation. Sometimes defective gene regulation is caused by mutations, sometimes by epimutations. Epimutations are often mistaken for mutations, especially when they affect oncogenes or tumor suppressor genes. Many cancers exhibit defective regulation of oncogenes and/or tumor suppressor genes, even when these genes have not mutated.[24] These nonmutation-caused alterations in the regulation of these genes are now known to be epigenetic.

On the epigenetic view, cancer is initiated by epigenetic disruptions such as global reductions in methylation, which are often present before any known mutations in oncogenes, including benign growths that precede cancer. The hypomethylation causes the instability in the chromosomes, emphasized by the aneuploid theory, as well as an increase in oncogene expression. Subsequently, there are specific changes in the methylation of particular genes. Oncogenes are further demethylated, while tumor suppressor genes are hypermethylated, thereby suppressing the tumor suppressors.

The progression of cancer often involves mutations and further alterations in chromosomal arrangements as well, but these genetic changes are considered secondary to the initial epigenetic changes, from this perspective. Moreover, epigenetic alterations also play an important role in the progression of cancer. That is, the progression itself is both genetic and epigenetic. This is true even for colorectal cancer, the poster case for the somatic mutation theory. As described earlier, each stage of colorectal cancer is accompanied by a new mutation. But no specific mutation can be linked to each stage. There is no recurrent mutation that can be said to cause the invasive properties of this cancer or its metastasis in all or even the majority of cases.[25] On the other hand, these properties have been linked to specific changes in the regulation of particular genes.

Some of the most compelling evidence for the primacy of epigenetics in cancer comes from a study on leukemia. As mentioned earlier, leukemia cells are highly aneuploid and mutated. Yet these cells can still be normalized through epigenetic interventions.[26] In this way, former leukemia cells can be made to behave like normal white blood cells. What is particularly noteworthy is that this normalization occurs without reversing the chromosomal rearrangements, which traditionally have been thought to cause the leukemia. The cells are still genetically abnormal, but they behave like normal white blood cells.

One prominent version of the epigenetic view touts the stem cell theory of cancer dynamics.[27] Others though, are compatible with the dedifferentiation perspective. In either case, a carcinogen is something that alters epigenetic regulation, which broadens that category considerably relative to somatic mutation theory. The therapeutic implications are also strikingly different because epigenetic processes, unlike genetic processes, are reversible, as the leukemia example dramatically demonstrates. There are also more ways to intervene epigenetically, and research in the development of epigenetic therapies is booming.[28] One potential advantage of epigenetic therapies over most therapies used currently is that they can be much more fine-tuned, compromising fewer healthy cells.

The epigenetic approach has some interesting implications for the T-devil as well. One tactic under discussion in the fight to save T-devils is a vaccine for DFTD. The problem with any vaccine is the evolution of genetic variants that allow the target of the vaccine to elude it. There is as yet little evidence of genetic variation in DFTD cells, but some researchers have begun to wonder about epigenetic variation.[29] This would present an even bigger problem than resistance based on standard genetic evolution, because epigenetic evolution is potentially much faster. It will be particularly interesting to know whether or how epigenetic variants function in evading the immune response of T-devils and the normalizing effects of the host tissue.

The Cancer Microenvironment

Genetic and most epigenetic theories of cancer are concerned primarily with what goes on inside the cancer cell. Recently, however, more attention is being paid to the microenvironment of the cancer cell. There are several distinct aspects of this environment, including the immune system, blood supply, and the normal tissue from which the cancer cells are derived—all of which have become important research areas. Collectively, these microenvironmental approaches take us furthest from SMT. They invite us to zoom out from the interior of the cancer cell to the surrounding tissue. It is only from this perspective that we can understand certain aspects of cancer behavior, not least of which is spontaneous remission.

Here I will focus on one of the microenvironmental approaches, called the tissue-based theory of cancer, according to which, cancer is the result of a breakdown in normal cell-cell interactions.[30] Call it a failure to communicate. The tissue-based theory of cancer complements and extends the epigenetic approach in important ways. First, it provides a mechanism for the initial epigenetic alterations, such as demethylation, that occur early in cancer development. Second, it provides a framework for understanding the genetic and epigenetic alterations that occur during cancer's progression. From this perspective, cancer's internal dynamic is largely a function of the normal cells from which cancer cells are derived and subsequently interact. These interactions can spur cancer development, or they can arrest its development and even eliminate every last vestige of a cancer. I have already described an example of the latter in Chapter 10.

Recall the study in which malignant melanoma cells were normalized by a microenvironment of embryonic stem cells. This is mysterious from the perspective of SMT; it is, however, not at all mysterious from the tissue-based perspective, but rather well within the range of normal cancer behavior. However, the embryonic stem cell environment is special in many ways. It is especially noteworthy, therefore, that other studies have found that cancers can be normalized by fully differentiated tissues.

Mary Bissell and her colleagues at University of California, Berkeley, constructed an artificial breast-tissue environment that simulated the essential qualities of normal breast tissue in three dimensions. They then introduced malignant breast cancer cells into this environment and waited to see what happened. The result came as a surprise to many, though not to Bissell: the cancer cells were normalized.[31] They lost their cancerous nature, in part through interactions with normal breast cells, arranged with the normal tissue architecture. But another important factor was the chemical composition of the extracellular matrix, the gel in which all cells are immersed. This gel is one of the primary ways through which cells chemically interact with each other during both normal development and cancer.

It is worth noting that Bissell came to cancer research with a background in developmental biology, and hence knowledge of the sort of cellular interactions that comprise normal development. For Bissell and other advocates of the tissue-based theory, cancer should be understood as a disruption of normal development, a disruption which, in some cases, self-corrects. This self-correction can occur in either the stem cell environment or in fully differentiated tissue.

Cancer, from this microenvironmentalist view, results from the disruption of normal interactions between cells. The disruption of cellular interactions alters the internal environment of the cells, which results in hypomethylation and other epigenetic changes. A carcinogen, on this view, is carcinogenic by virtue of disrupting normal cellular interactions within a tissue. Cancer development can potentially be detected much earlier on this view than it could on the SMT account, simply by monitoring the tissue architecture. Moreover, cancer therapies should focus more on helping the normal tissue cope with the cancer, the opposite of what occurs as a result of radiation and most forms of chemotherapy.

The tissue-based theory of cancer may also help shed some light on the T-devil cancer. From this perspective, DFTD represents a special challenge to normalization. Before this cancer evolved its transmissibility, it had to escape the normalizing influence of the normal tissue of the T-devil in which it initially evolved. That is a prerequisite for its metastasis. DFTD was then transmitted in this metastatic state. Metastatic cancer cells lack the organization of earlier-stage cancers; in fact, each cell is more like an individual organism. As such, DFTD cells are even immune to the influence of other DFTD cells. They are truly free agents and must be dealt with individually by the normal tissue in the T-devil's face and mouth.

Obviously, the fewer DFTD cells the infected T-devil has to deal with, the better. But even a relative few DFTD cells pose a problem for normalization because they are derived from different tissue than the host cells. It's as if they speak a different dialect, if not language, than the host cells. As such, it is more difficult for the normal cells to rein them in. That is true of metastatic cancer cells generally. But even metastatic cells can be normalized under the right conditions.

Should Father Damien Be a Saint?

By my lights, Father Damien was certainly saintly. He lived among and ministered to the lepers on the island of Molokai, until he himself succumbed to the disease. But that alone does not qualify him for sainthood according to the Catholic Church, which has some rather stringent criteria in that regard. One criterion seems particularly onerous from a scientific perspective: to be a true saint, you need to be responsible for two verified miracles. To make things a little easier, you can perform the miracles after you are dead. That is how Father Damien passed the threshold.

Being held responsible for curing advanced cancers has become a popular route to sainthood. Father Damien is only the latest example. A Hawaiian woman, Audrey Toguchi, went to Father Damien's gravesite and prayed for him to intercede with the divine to cure her of metastasized cancer. Her prayers, it seemed to her and the Catholic Church, were answered; she was soon cancer-free. Her doctor was as surprised as anyone. The Catholic Church determined that this remission could not have occurred without Father Damien's miraculous intercession. He was officially deemed a saint.

If the somatic mutation theory of cancer were gospel, the Catholic Church would be on solid ground. For a metastasized cancer to disappear, given SMT, it would have to undergo a series of extremely improbable reverse mutations. The case for sainthood becomes much weaker, though, if viewed from the epigenetic and especially microenvironmental perspectives. For example, the immune system might have come to the rescue in the nick of time, much as it seems to do in dogs with CTVT. I have also outlined another way that this woman could have been cured. Cancer, even in its most advanced stages, can be normalized through interactions with normal cells, whether in a stem cell environment or in that of fully differentiated tissue.

Spontaneous remissions of advanced cancers can occur without saintly intervention—even in atheists. Though unusual, this is well within the range of normal cancer behavior, at least from a microenvironmental perspective. That spontaneous remission can be explained without recourse to saints is a problem for the Catholic Church. That spontaneous remission seems miraculous from the perspective of the somatic mutation theory is a problem for the somatic mutation theory.

The microenvironmental perspective also provides some hope for the Tasmanian devil, even should the vaccine prove unworkable. Drugs that boost the immune response, as well as those that promote normalization, could be helpful. But the best hope for T-devils is for them to naturally develop immune or normalizing responses to this cancer much as domestic dogs have done to CTVT. In the meantime, it wouldn't hurt to pray for the devil.

Devils and Saints

The T-devil's cancer is both typical and exceptional. It is typical at the cellular level, in being poorly differentiated and aneuploid. And though we can't know for sure, there is no reason to believe it had an atypical origin, whether from stem cells or fully differentiated tissue. Nor is there reason to assume that the mechanism underlying its cancerous transformation and progression required different processes than other cancers. But it is these processes that are most in dispute.

The SMT account is based on the executive view of genes as cellular directors, with an emphasis on oncogenes and tumor suppressor genes. The aneuploid theory also stems from the executive gene view, but with a different set of genetic actors as initiators of the cancerous transformation, then shifts to a more macroscopic chromosomal perspective for the further transformations during cancer development. On the epigenetic account, both genetic theories have missed the first step in cancer development, the step before any mutations occur, which is a reversible epigenetic event. From this perspective, even very advanced cancers are also epigenetically reversible under the right conditions. The microenvironmental perspective, which includes tissue-based theories, supplements the epigenetic approach in demonstrating what those right conditions are. The epigenetic and microenvironmental models of cancer are more compatible with the executive cell perspective.

What is exceptional, though not unique, in the T-devil's cancer is its transmissibility. We can consider this the next stage after metastasis, a stage which, fortunately, most cancers never get to. Transmissibility appears to require a degree of stability at the cellular level that most cancers aren't able to achieve. Also required are evasion of the immune response and the normalizing influence of the host tissue—microenvironmental factors both. Given the power of cancer microenvironments to either eliminate or epigenetically reverse even the most advanced cancers, there is reason for hope, even for those without recourse to saints.

Postscript

The Janus Gene

IN THIS SHORT TOUR, I HAVE HAD TIME TO COVER ONLY A FEW highlights of this exciting new science known as epigenetics. Here I want to briefly reprise a few important themes that have emerged during the course of the tour.

The first theme concerns the nature of epigenetic processes: a form of gene regulation. Epigenetic gene regulation is long-term gene regulation, hence epigenetic alterations have long-term effects on gene behavior. Indeed, epigenetic alterations of gene behavior can be longer lasting than mutational alterations of gene behavior. But unlike the alterations of gene behavior caused by mutations, epigenetic alterations of gene behavior are generally reversible.

The second theme is that our environment affects the behavior of our genes, both in the short and the long term. The long-term environmental influences on genes' behavior come by way of epigenetic processes. Environmentally induced epigenetic alterations that occur early in our lives are especially important. We have explored, in particular, the epigenetic effects of poor nutrition and stress on the fetus and the infant, and their myriad health consequences during adulthood. But our environment continues to epigenetically influence our genes throughout our lives.

The third theme is randomness. Epigenetic processes, like all biological processes, have a random element, and sometimes this random element looms large. This is true, for example, of methylation at the agouti locus, which affects not only coloration but susceptibility to obesity, diabetes, and cancer in mice. X-chromosome inactivation is another epigenetic process for which randomness is critical. Indeed, in this case, we could say that randomness is adaptive. Without it there certainly wouldn't be any X-women.

Clones, whether natural as in monozygotic twins, or manufactured as in Cc the calico cat, are far from carbon copies. There are a number of reasons for this, some of them epigenetic. In the case of Cc, random X-chromosome inactivation caused her coloration to diverge markedly from that of her mother, to the point that she even lacked one pigment entirely.

Both random and environmentally induced epigenetic differences are evident in monozygotic twins. We began this book with a particularly dramatic case of epigenetic discordance for Kallmann syndrome in human clones. Other examples of clone discordance include Alzheimer's disease, lupus, cancer, and color discrimination.

Some epigenetic alterations of gene behavior have effects that extend beyond an individual lifetime. This is theme number four. The effect of these transgenerational epigenetic alterations may be direct or indirect. Direct transgenerational effects occur when the epigenetic mark is transmitted directly from parent to offspring, through sperm or egg. This is what I call "true epigenetic inheritance." True epigenetic inheritance is not common in mammals like us, but it does occur. Indirect transgenerational effects are much more common.

The most direct of these indirect transgenerational epigenetic effects is genomic imprinting, in which the original epigenetic mark in the parent is reproduced with great fidelity in the offspring. Much more indirect are the transgenerational effects observed in the maternal behavior and stress response of rats. Here, the epigenetic alterations that influence these behaviors are recreated through the social interactions that they both influence and are influenced by. This transgenerational effect is a positive feedback loop involving gene action and social interaction. Whether direct or indirect, these transgenerational epigenetic effects should expand our notion of inheritance.

The fifth and final theme is actually a meta-theme, a theme comprised of the previous four themes. This meta-theme concerns some of our basic intuitions about the role of genes in explaining biological processes ranging from protein synthesis to cellular differentiation and cancer. Genes are traditionally viewed as biochemical executives that initiate and direct these processes, in contradistinction to all other biochemicals within a cell, which function in a more blue-collar way. I used the metaphor of the theatrical production, the play, to illustrate this view: Genes are the directors, proteins the actors, and all other biochemicals act as stagehands. From an alternative perspective, advocated here, this play is more improvisational and genes are more like members of an ensemble cast, a cast that includes proteins and other biochemical actors. Gene actions are as much effect as cause during protein synthesis, and genomic activity is as much effect as cause during cellular differentiation, both normal and pathological.

From this alternative perspective, genes have two faces, two aspects, like the Roman deity Janus, the god of doorways and gates, of entryways and exits, of beginnings and endings. Only one aspect, the outward-facing, causal aspect, is acknowledged on the traditional account. The result is a simplistic and distorted view of genes and gene actions. For genes also have another aspect, the inward-facing, responsive aspect. This responsive aspect of the Janus gene is highlighted in epigenetic research, the payoffs of which, even in these early days, have been enormous.

Acknowledgments

I benefited immensely from the insightful comments of Tamara Bushnik, Peter Godfrey-Smith, and Eva Jablonka, for which I am deeply grateful. Thanks also are due my agent, Lisa Adams, for her advice and encouragement throughout the book-writing process, and to my editor, Jack Repcheck, for his enthusiasm and guidance.

Notes

Preface. **What Your Genes Are Wearing**

1. Christian, Bixler, et al. (1971).

2. Schwanzel-Fukuda, Jorgenson, et al. (1992); Whitlock, Illing, et al. (2006).

3. See, for example, Bianco and Kaiser (2009) for a recent review of the genetics of Kallmann syndrome.

4. Hipkin, Casson, et al. (1990). French, Venu, et al. (2009) is another case study of twin discordance with respect to Kallmann syndrome, but this discordance is not as severe.

5. See, for example, Wong, Gottesman, et al. (2005).

6. This is a characterization of epigenetics in the narrow sense, which is the definition of epigenetics I will use throughout this book. Epigenetics in the broad sense is not confined to alterations in DNA. See Jablonka and Lamb (2002) for a good historical and conceptual overview of the term.

7. See the following regarding epigenetic sources of discordance: for lupus, Ballestar, Esteller, et al. (2006); and for Alzheimer's disease, Mastroeni, McKee, et al. (2009). Singh and O'Reilly (2009) provide evidence of epigenetic divergence in monozygotic twins discordant for schizophrenia; see also Kato, Iwamoto, et al. (2005).

Chapter 1. **A Grandmother Effect**

1. Stein and Susser (1975).

2. This ongoing longitudinal study is an international collaboration, involving several departments of the Academic Medical Center in Amsterdam and the MRC (Medical Research Council) at the University of Southampton in England.

3. Smith (1947).

4. Stein, Susser, et al. (1972); Ravelli, Stein, et al. (1976).

5. Hoch (1998). For the association between famine and depression, see Brown, van Os, et al. (2000); between famine and antisocial personality disorders in males, see Neugebauer, Hoek, et al. (1999).

6. Stein, Ravelli, et al. (1995); Lumey and Stein (1997); Lumey (1998).

7. Ravelli, van der Meulen, et al. (1998); Roseboom, van der Meulen, et al. (1999, 2000a, 2000b); Painter, Roseboom, et al. (2005).

8. Roseboom, de Rooij, et al. (2006).

9. Tobi, Lumey, et al. (2009).

10. Painter, Osmond, et al. (2008).

Chapter 2. **Directors, Actors, Stagehands**

1. Allen (1978) is an outstanding biography of Morgan from which this discussion benefited.

2. The original publications were Watson and Crick (1953a, 1953b). Watson subsequently wrote an account of this research—from his perspective—for a lay audience (Watson 1968), which is much more gossipy than most scientific memoirs of this sort.

3. The original formulation—of George Beadle and Edward Tatum—was actually "one gene = one enzyme" (Beadle and Tatum 1941). This was modified in the late 1950s to "one gene = one polypeptide (protein)" to include nonenzymatic proteins.

4. Posttranslational processing should actually be considered the third stage of protein synthesis since translation products are rarely functional. During posttranslational processing, many changes are made to the protoprotein to turn it into something that is physiologically useful. A good

example is the protoprotein templated by the proopiomelanocortin (POMC) gene. The POMC protoprotein is cleaved into one of 20 different smaller protein hormones, depending on the type of cell in which the protoprotein finds itself. In the cells of one lobe of the pituitary gland (a small endocrine gland at the base of the brain), POMC is cleaved into adrenocorticotropic hormone (ACTH), which functions in the stress response. In cells of another part of the pituitary, POMC is cleaved into the opiate β-endorphin. In skin cells, POMC is cleaved into melanocyte stimulating hormone, which promotes production of the black pigment melanin. The protein product of the POMC gene is not a simple function of its base sequence, but rather is determined at the cellular level.

Other sorts of posttranslational processing take us even further from the DNA base sequence. Sometimes the amino acid that was coded is chemically converted into another amino acid that was not coded. In other cases, one amino acid is substituted for another amino acid in the protoprotein. Sometimes the substitute amino acid is not even one of the 20 amino acids for which there is a code. In these cases, the relationship between the DNA base sequence and the protein amino acid sequence is altered at the cellular level.

5. The executive cell view advocated here has a long history in biology. Mention should be made of Ernest Everett Just (1883–1941) among the early proponents; see, for example, Just (1939). Just formulated an idea of gene action and reaction (cytoplasm-nucleus interactions) very much along the lines advocated here (see, for example, Sapp 2009 and Newman 2009).

6. Even in the first two stages of protein synthesis, the process is shaped in a cell-dependent manner. Stage 1, transcription (see the figure on page 18), is actually a two-step process. The first step is the formation of a pre–messenger RNA. The second step is the transformation of the pre-mRNA into the final mRNA, the mRNA that will actually serve as the template for protoprotein construction. A lot happens during this second step. Each gene actually consists of two or more separate coding regions called exons, separated by noncoding regions called introns. All of it—exons and introns—is transcribed into the pre-mRNA; then, following transcription, the intronic RNA bits are removed and the exons are spliced together. For many genes, the exons can be spliced together in different ways, each spliced variant constituting a different protein. This is called *alternative splicing*, which is another way that more than one protein can be constructed from one gene.

Alternative splicing is but one way in which the initial transcript is modified according to the cellular environment. After all of the splicing comes RNA editing. During the editing process, some individual bases in the mRNA sequence undergo a chemical conversion into a different base, one that was not encoded in the DNA. So, even the correspondence between the DNA base sequence and the mRNA sequence is not always one-to-one.

7. See, for example, McClellan and King (2010) and Galvan, Falvella, et al. (2010).

8. See, for example, Rakyan, Blewitt, et al. (2002), and Hatchwell and Greally (2007).

9. Griffiths and Neumann-Held (1999); Beurton, Falk, et al. (2000); Stoltz, Griffiths, et al. (2004); Rheinberger (2008); Portin (2009).

10. Genes that don't code for proteins are often referred to as "RNA genes."

11. The term *control panel* is purely metaphoric, intended to aid the intuitions of nonbiologists regarding garden-variety gene regulation. Moreover, the regulatory elements that comprise the control panel need not be physically contiguous. Finally, some or all of the regulatory elements that comprise a control panel for a particular gene may also function in the regulation of other genes.

Chapter 3. **What Roids Wrought**

1. Guerriero (2009) summarizes what is known about the distribution of androgen receptors in the brain.

2. For an example from fish studies, see Hannes, Franck, et al. (1984). Burmeister, Kailasanath, et al. (2007) reported that androgen receptor levels dropped in response to a drop in social status. The best evidence of this effect in humans comes from studies of athletes after competition. For tennis players, Booth, Shelley, et al. (1989) reported a drop in testosterone levels in those who lost a match and an elevation of testosterone levels in winners. But the effect of competition outcome (or aggressive interactions in other animals) on testosterone levels is actually quite complex. See, for example, Suay, Salvador, et al. (1999). I am simplifying things here.

3. More precisely, pituitary GT levels are controlled by a small population of neurons within the hypothalamus called the preoptic area (POA); see, for example, Francis, Soma, et al. (1993).

4. In the scientific literature, the acronym for gonadotropin releasing hormone is GnRH, not GTRH. I use GTRH partly because it is easier for the uninitiated to assimilate, partly because it is more consistent with the standard abbreviation for gonadotropin (GT).

5. For more detailed accounts of the relationship between dominance, testosterone, and GTRH levels, see Francis, Jacobson, et al. (1992).

6. Francis, Soma, et al. (1993).

7. White, Nguyen, et al. (2002). Initially, though, there is a transient increase in the activity of the GTRH gene in males who are making a downward social transition (Parikh, Clement, et al. 2006), so the effect of social status on the GTRH gene itself is complicated. It may well be that transcription of this gene is affected less than its translation. Not all messenger RNAs get translated to protoproteins; it is often the case that more mRNAs are made than proteins (in this case, GTRH). The excess mRNA gets degraded. In the case of the GTRH gene, the rate of this degradation may be related to social status. GTRH mRNA may be degraded at a higher rate in nonterritorial males than in territorial males.

8. Renn, Aubin-Horth, et al. (2008).

9. For change in androgen receptor gene, see Burmeister, Kailasanath, et al. (2007); for change in GTRH receptor gene, see Au, Greenwood, et al. (2006).

Chapter 4. **The Well-Socialized Gene**

1. After viewing *The Deer Hunter* in March, 1979, Jan Scruggs, a Vietnam veteran, hit on the idea of a memorial with the names of all who were killed during the conflict (Ashabranner 1988).

2. Glucocorticoids such as cortisol, unlike testosterone, also exert effects through two nongenomic pathways—first by way of protein-protein interactions with other transcription factors (see, for example, Revollo and Cidlowski 2009), and second by way of nonnuclear receptors. This latter pathway is thought to underlie the most rapid responses to these hormones (Evanson, Tasker, et al. 2010).

3. Another good indicator of a pathological stress response is the neuropeptide arginine vasopressin (AVP); see, for example, Lightman (2008). Both CRH and AVP increase rapidly in response to an acute stressor, though

chronic stress often results in a reduction in CRH over time. AVP increases with chronic stress, however.

4. Ventolini, Neiger, et al. (2008); Bevilacqua, Brunelli, et al. (2010).

5. Seckl and Holmes (2007); Drake, Tang, et al. (2007).

6. Kapoor, Leen, et al. (2008); Seckl (2008).

7. See, for example, Seckl and Meaney (2006), and Kapoor, Petropoulos, et al. (2008).

8. See *PTSD Forum: Promoting Growth Through Healing,* http://www.ptsd forum.org).

9. Laugharne, Janca, et al. (2007).

10. Yehuda, Bell, et al. (2008); Yehuda and Bierer (2007).

11. Yehuda, Engel, et al. (2005); Brand, Engel, et al. (2006).

12. Dean, Yu, et al. (2001). The effect of synthetic glucocorticoids is highly sex-specific and depends critically on the timing of the exposure (Kapoor and Matthews 2008; Kapoor, Kostaki, et al. 2009).

13. Kapoor, Petropoulos, et al. (2008); Emack, Kostaki, et al. (2008).

14. Liu, Diorio, et al. (1997); Francis and Meaney (1999); Francis, Champagne, et al. (2000).

15. Francis, Diorio, et al. (1999).

16. Liu, Diorio, et al. (1997).

17. Francis, Champagne, et al. (1999); Szyf, Weaver, et al. (2005).

18. Weaver, Cervoni, et al. (2004).

19. The methyl group does not attach just anywhere on the DNA, but rather specifically to a cytosine that is adjacent to a guanine. Since all of the bases in the DNA sequence are separated by a phosphorus molecule, the convention is to label these sites "CpG."

20. Goldberg, True, et al. (1990); Kaminsky, Petronis, et al. (2008); Coventry, Medland, et al. (2009).

Chapter 5. **Kentucky Fried Chicken in Bangkok**

1. Neel (1962).

2. See, for example, Rothwell and Stock (1981), Speakman (2006, 2008), and Gibson (2007).

3. Neel (1999) abandoned the thrifty-genes hypothesis. Newer versions include Prentice, Hennig, et al. (2008), and Wells (2009).

4. See, for example, Hinney, Vogel, et al. (2010).

5. In this respect, obesity resembles many other complex traits (Petronis 2001; Smithies 2005).

6. For treatments of the story about the complexity of obesity genes, see Shuldiner and Munir (2003); Damcott, Sack, et al. (2003); Swarbrick and Vaisse (2003).

7. De Boo and Harding (2006) is a good summary of the diseases linked to birth weight.

8. Warner and Ozanne (2010). This is also called the *developmental origins* hypothesis to distinguish it from gene-centered theories (Mcmillen and Robinson 2005; Waterland and Michels 2007).

9. See, for example, Barker, Robinson, et al. (1997), and Hales and Barker (2001).

10. See, for example, Susser and Levin (1999).

11. For reviews, see Junien and Nathanielsz (2007) and Burdge, Hanson, et al. (2007).

12. Seckl (2004); Seckl and Holmes (2007).

13. Lillycrop, Slater-Jefferies, et al. (2007); Kim, Friso, et al. (2009). There are a number of DNA methyltransferases (Dnmt); the discussion in the text that follows concerns Dnmt1.

14. Bellinger and Langley-Evans (2005); Lillycrop, Phillips, et al. (2005).

15. Lillycrop, Slater-Jefferies, et al. (2007).

16. It is not only glucocorticoid receptor gene (*GR*) expression in the liver that influences these conditions. For example, *GR* expression in adipose tissue (fat) also plays an important role in the metabolic syndrome (see, for example, Walker and Andrew 2006). For an example of how *GR* expression in the liver is related to diabetes, see Simmons (2007). For a general overview of the relationship between *GR* expression and the metabolic syndrome, see Witchel and DeFranco (2006). And for a good review of tissue-specific glucocorticoid action, see Gross and Cidlowski (2008).

17. Meaney, Szyf, et al. (2007).

18. Shively, Register, et al. (2009).

19. Bjorntorp and Rosmond (2000); Taylor and Poston (2007).

20. There are a number of histones, divided into five classes, which have combinatorial properties analogous to the bases in the genetic code, and hence comprise a "histone code" (Strahl and Allis 2000). This histone code would be an epigenetic code. But I don't find this code talk helpful.

21. Aagaard-Tillery, Grove, et al. (2008); Delage and Dashwood (2008).

22. In the methylation of DNA, the methyl group is bonded to a cytosine base; in histone methylation, the methyl group is bonded to an amino acid, usually a lysine or arginine. As in DNA methylation usually (but not always), histone methylation has a suppressive effect on gene expression. There are a number of other sorts of posttranslational modifications of histones with epigenetic consequences, including acetylation, in which an acetyl group (CH_3CO) is added to the lysine of the histone. Acetylation of histones usually (but not always) promotes gene expression.

23. Lillycrop, Slater-Jefferies, et al. (2007).

24. See, for example, Simmons (2007); Hess (2009); Zeisel (2009).

25. Kim, Friso, et al. (2009).

26. Rogers (2008); Leeming and Lucock (2009).

27. Jones, Skinner, et al. (2008); Currenti (2010); Ptak and Petronis (2010).

Chapter 6. Twigs, Trees, and Fruits

1. Beck and Power (1988); Porton and Niebrugge (2002). Not surprisingly, such sexually incompetent males also have reduced reproductive success (Meder 1993; Ryan, Thompson, et al. 2002). It is worse for hand-raised male chimps, however, almost half of which (46 percent) fail to exhibit "appropriate sexual behavior" (King and Mellen 1994).

2. I trace all work on social inheritance to the pioneering research of Denenberg and Rosenberg (1967) who demonstrated that manipulations of female rats as pups affected the emotionality (as open field activity, which is roughly a behavioral stress measure) and weight of their grand-offspring. This is the first grandmother effect that I am aware of. I believe this work went largely unappreciated until Michael Meaney and his colleagues recognized its importance.

3. Harlow and Zimmerman (1959).

4. Harlow, Harlow, et al. (1971).

5. Ruppenthal, Arling, et al. (1976); Champoux, Byrne, et al. (1992).

6. Champagne and Meaney (2001).

7. Champagne, Weaver, et al. (2006); Champagne and Curley (2009).

8. Champagne, Diorio, et al. (2001); Ross and Young (2009).

9. Champagne, Weaver, et al. (2006).

10. Bardi and Huffman (2006); McCormack, Sanchez, et al. (2006).

11. For maternal effects on Japanese macaques, see Bardi and Huffman (2002, 2006); for maternal effects on pigtail macaques, see Weaver, Richardson, et al. (2004).

12. Maestripieri (2003, 2005).

13. Greenfield and Marks (2010).

14. Bradley, Binder, et al. (2008).

15. Serbin and Karp (2004); Bailey, Hill, et al. (2009).

16. McGowan, Sasaki, et al. (2009). See also Weaver (2009).

17. Patton, Coffey, et al. (2001); DiBartolo and Helt (2007); Otani, Suzuki, et al. (2009). See Joyce, Williamson, et al. (2007), for effects of affectionless control on the stress axis.

18. Engert, Joober, et al. (2009); Kochanska, Barry, et al. (2009); Kaitz, Maytal, et al. (2010).

19. See, for example, Calatayud and Belzung (2001), Champagne and Meaney (2001), and Weaver (2009).

20. See, for example, Calatayud and Belzung (2001), and Champagne and Curley (2009).

21. Tyrka, Wier, et al. (2008).

22. Weaver, Meaney, et al. (2006).

23. Weaver, Champagne, et al. (2005).

Chapter 7. **What Wright Wrought**

1. Castle, Carpenter, et al. (1906). For a brief biography of Castle's scientific life, see Snell and Reed (1993).

2. Provine (1986) is an excellent biography of Wright, with an emphasis on his contributions to population genetics and evolutionary biology.

3. Castle and Wright (1916); Wright (1916, 1927).

4. Voisey and van Daal (2002) is a detailed account of the physiological actions (at the molecular level) of the agouti protein and its regulation.

5. Wilson, Ollmann, et al. (1995).

6. Miltenberger, Mynatt, et al. (1997); Morgan, Sutherland, et al. (1999). The *viable yellow* mutation actually occurs somewhat upstream of the agouti locus itself. A mobile genetic element called a retrotransposon drives so-called ectopic expression of this gene (Duhl, Stevens, et al. 1994; Duhl,

Vrieling, et al. 1994). The specific type of retrotransposon in this case is an intracisternal A particle (IAP). IAPs are involved in other dominant mutations at this locus as well. It is the IAP that is methylated, not the control panel of the agouti allele.

7. Wolff, Roberts, et al. (1986).

8. Wolff (1996).

9. Michaud, van Vugt, et al. (1994).

10. Morgan, Sutherland, et al. (1999).

11. Martin, Cropley, et al. (2008).

12. Morgan, Sutherland, et al. (1999).

13. Wolff, Kodell, et al. (1998); Dolinoy, Weidman, et al. (2006).

14. Cropley, Suter, et al. (2006). But see Waterland, Travisano, et al. (2007) for a different interpretation of these results. Blewitt, Vickaryous, et al. (2006) provides evidence that the methylation state itself is not the inherited epigenetic state in these experiments.

15. Rakyan, Preis, et al. (2001); Waterland, Travisano, et al. (2007).

16. See, for example, Reik, Dean, et al. (2001).

17. Rakyan, Chong, et al. (2003). Interestingly the $Axin^{fu}$ mutation, like the A^{vy} mutation, involves an IAP (see Note 6).

18. Reviewed in Roemer, Reik, et al. (1997). Two of the best examples are studied in Rassoulzadegan, Grandjean, et al. (2006) and Rassoulzadegan, Grandjean, et al. (2007) involving *Kit*, another locus involved in coat coloration. This form of epigenetic inheritance appears to involve an RNA-based form of epigenetic regulation that I will discuss later.

19. Martin, Ward, et al. (2005); Morak, Schackert, et al. (2008). But see Chong, Youngson, et al. (2007).

20. Jablonka and Raz (2009). One reason that epigenetic inheritance is much more common in plants (and fungi) is that they don't exhibit the early segregation of the germ line characteristic of muticellular animals. Jablonka and Raz further speculate that epigenetic inheritance is more adaptive in plants and fungi because they lack complex nervous systems and hence behavioral plasticity. These authors further speculate that epigenetic inheritance is actively selected against in highly mobile animals because they experience less-predictable environments, and as such there is less correlation between the environments of parent, offspring, and grand-offspring.

21. See Jablonka and Raz (2009) for a thorough review.

22. Richards (2006) and Henderson and Jacobsen (2007) are excellent reviews of epigenetic inheritance in plants.

23. Stokes, Kunkel, et al. (2002); Stokes and Richards (2002). This RNA-based form of epigenetic inheritance (see also Note 18) is often referred to as a paramutation, in which an epiallele in one generation affects the expression of the other allele at that locus in the next generation.

24. Koornneef, Hanhart, et al. (1991).

25. Zilberman and Henikoff (2005).

26. Here I am adopting the terminology of Youngson and Whitelaw (2008).

27. Moreover, grandsons of those who experienced plentiful food were more susceptible to diabetes (Pembrey, Bygren, et al. 2006). For a discussion of the role of sperm epigenetics in development, see Carrell and Hammoud (2010). For a mechanism of epigenetic inheritance via histones and chromatin remodeling, see Puri, Dhawan, et al. (2010).

Chapter 8. **X-Women**

1. As Dobyns, Filauro, et al. (2004) emphasize, most X-linked traits are neither dominant nor recessive but exhibit variable penetrance. There is evidence for such variability in color blindness as well.

2. Kraemer (2000). There are, of course, many other factors for this higher male mortality risk than just the male X deficit. For an interesting sociological study of how men and women tend to explain these differences, see Emslie and Hunt (2008).

3. This pioneering work in neurobiological genetics was conducted by Jeremy Nathans and his coworkers (Nathans, Piantanida, et al. 1986; Nathans, Thomas, et al. 1986). See also Nathans (1999).

4. Jordan and Mollon (1993); Jameson, Highnote, et al. (2001).

5. X-chromosome epigenetics can be traced to Mary Lyon's discovery of X inactivation (Lyon 1961); see also Lyon (1971) and Lyon (1989). Susumu Ohno, a giant in the field of genetics, especially in sex chromosome research, was the first to propose methylation as a mechanism for inactivation (Ohno 1969). The important contribution of Ohno to this field is summarized in Riggs (2002). Lyon (1995) is a historical overview of research on X-chromosome inactivation. Chow, Yen, et al. (2005) is a

good overview of what is known of the epigenetics of X inactivation. Urnov and Wolffe (2001) is a good history of epigenetics that covers the role of X inactivation in the development of the field (see also Holliday 2006). Jablonka (2004) provides an evolutionary perspective on the epigenetics of X inactivation.

6. Lyon (1961).

7. Brown and Greally (2003); Berletch, Yang, et al. (2010).

8. Namekawa, VandeBerg, et al. (2007); Deakin, Chaumeil, et al. (2009). There is some expression of paternal X genes in some tissues (VandeBerg, Johnston, et al. 1983).

9. Some features of autosomes appear to prevent complete inactivation; see, for example, Popova, Tada, et al. (2006).

10. Most dramatically, Cc completely lacked any of the orange coloration of Rainbow, indicating that an X-linked gene involved in the production of orange fur was randomly turned off early in Cc's development.

11. This is certainly true of mice (Wagschal and Feil 2006); it is less clearly the case in humans (Moreira de Mello, de Araujo, et al. 2010).

12. Erwin and Lee (2008).

13. There is some evidence for this (Tiberio 1994; Loat, Asbury, et al. 2004; Haque, Gottesman, et al. 2009). Moreover, there is a report of female monozygotic twins discordant for red-green color deficiency who exhibit different X-inactivation patterns in their cone cells (Jorgensen, Philip, et al. 1992).

14. Pardo, Pérez, et al. (2007); Rodriguez-Carmona, Sharpe, et al. (2008).

15. Verriest and Gonella (1972); Cohn, Emmerich, et al. (1989).

16. Deeb (2005) is a nice summary of the molecular biology discussed here. See also Hayashi, Motulsky, et al. (1999).

17. Jordan and Mollon (1993).

18. Hunt, Williams, et al. (1993); Shyue, Hewett-Emmett, et al. (1995).

19. Tovee (1993).

20. Jacobs (1998, 2008); Jacobs and Deegan (2003).

Chapter 9. Horses Asses

1. I got this information online from *The Mule Page*, http://www.phud pucker.com/mules/mule.htm.

2. *The Reivers* (1962).

3. There is a characteristic cognitive deficit known as Turner neurocognitive phenotype (Ross, Roeltgen, et al. 2006), which is primarily restricted to spatial and mathematical reasoning. Turner syndrome is also associated with autism.

4. Parent-of-origin effects in Turner syndrome have been reported for growth (Hamelin, Anglin, et al. 2006; but see Ko, Kim, et al. 2010) and for cognition (see, for example, Skuse, James, et al. 1997, and Crespi 2008).

5. Cassidy and Ledbetter (1989).

6. Chen, Visootsak, et al. (2007).

7. Driscoll, Waters, et al. (1992); Williams, Angelman, et al. (1995).

8. Bittel, Kibiryeva, et al. (2005). This state of permanent paternal X inactivation is referred to as uniparental disomy.

9. Weksberg and Squire (1996); Delaval, Wagschal, et al. (2006).

10. Weksberg, Shuman, et al. (2005). Wilms' tumor is an embryonic cancer. Embryonic cancers are rare, generally occurring only when there are major developmental problems.

11. See, for example, Reik (1989) and Shire (1989).

12. Reik, Dean, et al. (2001).

13. Santos and Dean (2004).

14. There is increasing evidence of defects in the reprogramming of imprinted genes in embryos generated through assisted reproductive technologies (ARTs); see, for example, Grace and Sinclair (2009), Laprise (2009), and Owen and Segars (2009). Defective reprogramming of imprinted genes is also thought to explain why mammal cloning is so hard to accomplish.

15. Lewis and Reik (2006).

16. When paternally imprinted genes are overexpressed, the result is often an abnormally large placenta; see, for example, Reik, Constancia, et al. (2003), and Fowden, Sibley, et al. (2006).

17. Overexpression of paternally imprinted genes often results in fetal overgrowth (Cattanach, Beechey, et al. 2006; Biliya and Bulla 2010).

18. The inhibitor of IGF2 under discussion here is called H19, which is an untranslated mRNA. The regulation of IGF2 is quite complex, and involves other loci and alleles.

19. Uniparental disomy of this sort is responsible for about 20 percent of the cases of Beckwith-Wiedemann syndrome (Cooper, Curley, et al. 2007).

20. See, for example, Bartholdi, Krajewska-Walasek, et al. (2009).

21. Kinoshita, Ikeda, et al. (2008).

22. The syndromes discussed here are but a tiny sample of what goes wrong health-wise when imprinting goes awry. See Amor and Halliday (2008) for a review of imprinting-related disorders. Wadhwa, Buss, et al. (2009) discuss imprinting disorders in the context of epigenetics and disease generally. Murphy and Jirtle (2003) discuss the costs of monoallelic expression in an evolutionary context.

23. See, for example, Vos, Dybing, et al. (2000), and Hayes, Stuart, et al. (2006).

24. For a discussion of the sexual plasticity of fishes relative to that of other vertebrates, see Francis (1992).

25. Gross-Sorokin, Roast, et al. (2006). Intersex males—that is, genetic males with ovarian features—are also common (Jobling, Williams, et al. 2006).

26. Milnes, Bermudez, et al. (2006).

27. Crews (2010) provides an excellent overview of endocrine disruptors and imprinted genes. See also Prins (2008) and Skinner, Manikkam, et al. (2010).

28. Virtanen, Rajpert-De Meyts, et al. (2005); Diamanti-Kandarakis, Bourguignon, et al. (2009); Wohlfahrt-Veje, Main, et al. (2009); Soto and Sonnenschein (2010). Some of these later-developing syndromes come by way of the agouti locus, which is maternally imprinted (Morgan, Sutherland, et al. 1999). Bisphenol A (BPA) shifts the coat color of A^{vy} toward the yellow (and hence unhealthy) part of the spectrum through its hypomethylating effects (Dolinoy, Huang, et al. 2007). Interestingly, a high-folate diet reverses this effect. Bernal and Jirtle (2010) warn that BPA exposure could have significant health consequences for humans, both in this generation and, through epigenetic inheritance, future generations.

29. Anway, Cupp, et al. (2005).

30. Chang, Anway, et al. (2006); Stouder and Paoloni-Giacobino (2010).

31. Anway and Skinner (2008).

32. Shi, Krella, et al. (2005).

Chapter 10.   Sea Urchins Are Not Just to Eat

1. Monroy (1986) is an excellent overview of the importance of the sea urchin in developmental biology.

2. See Bodemer (1964) for the early history of preformationism. Caspar Friedrich Wolf (1733–1794), considered one of the founders of embryology, decisively refuted the early forms of preformationism.

3. For more detailed discussions regarding the differences between preformationism and epigenesis, see Van Speybroeck, De Waele, et al. (2002), and Maienschein (2008).

4. This version of preformationism was first advocated by August Weismann and is known as the mosaic theory of development (see the references cited in Note 3).

5. Scott Gilbert provides a nice account of these experiments in his *Developmental Biology* (pp. 287–289, in the 3rd ed., 1991).

6. For my purposes here, self-organizing processes have two features: first, the relevant elements act in parallel (simultaneously) rather than serially; second, the executive function is distributed, not centralized. See ten Berge, Koole, et al. (2009) for a good example of self-organization during early development. In that study, the investigators focused on the activity of the gene *Wnt* and demonstrated the effects of undirected cellular interactions on its expression.

7. For this reason, Driesch could be considered the first proponent of the executive cell view advocated in this book.

8. Spooky notions like entelechy are generally subsumed under the term "vitalism" by biologists. Vitalism should be distinguished from organicism, which is a thoroughly materialist (nonspooky) but nonreductionist framework for understanding development and other complex phenomena. By nonreductionist, I mean a rejection of the notion that an explanation confined to a characterization of the properties of the parts is sufficient for explaining the whole. Put another way, reductionist explanations are entirely bottom-up, while nonreductionist explanations are both bottom-up and top-down.

Important organicist developmental biologists include Karl Ernst von Baer (1792–1876), Charles Otis Whitman (1842–1910), Oskar Hertwig (1849-1922), Hans Spemann (1869–1941), Ross Granville Harrison (1870–1959), Ernest Everett Just (1883–1941), Paul Alfred Weiss (1898–1989), Viktor Hamburger (1900–2001), Joseph Needham (1900–1995), and Conrad Waddington (1905–1975). Scott Gilbert is a prominent contemporary organicist; see Gilbert and Sarkar (2000) for an excellent history of organicism. For a good account of contemporary organicism (under another name), see Kirschner, Gerhart, et al. (2000). Organicists reject the machine analogy for biological systems. Organicist developmental biologists reject preformationism.

9. Epigenesists do not deny the importance of these initial conditions (particularly the genome) in determining the adult form, only that these initial conditions do not *comprise* the adult form, however latently.

10. In contradistinction to the executive genome view, the executive cell perspective that I advocate falls into the category of organicist versions of epigenesis.

11. Ultimately, I would suggest, the directorial intuition in preformationism (and creationism) reflects a deep anthropomorphism that stems from the way we relate to our artifacts. I believe that this anthropomorphism is a major impediment to understanding complex processes such as development (and evolution).

12. This point is made especially well in Susan Oyama's *The Ontogeny of Information* (1985). For insightful critiques of the "genome as recipe/ program" metaphors, see Nijhout (1990), Atlan and Koppel (1990), Moss (1992), Fox Keller (1999, 2000), and Pigliucci (2010). Both Atlan and Koppel, and Fox Keller, note that there is a problem with the data-program distinction as well as the software-hardware distinction noted here. Nijhout advocates treating genes as material resources for the developing organism, as advocated here. This way of thinking about genes and genomes is also characteristic of advocates for developmental systems theory (I prefer the term "developmental systems perspective"). See, for example, Oyama (1985), and Griffiths and Gray (1994).

13. Here I am ignoring the fact that a small percentage of cells acquire slightly different genomes through somatic mutations. These somatic mutations do not, however, figure in normal cellular differentiation. I should also note that red blood cells do not have genomes when mature.

14. The canonical microRNA is *lin-4*, first identified in the nematode *Caenorhabditis elegans* (Horvitz and Sulston 1980). It is 22 nucleotides in length, with a hairpin structure characteristic of microRNAs. For good reviews, see Eddy (2001) and Storz, Altuvia, et al. (2005); see also Ying, Chang, et al. (2008).

15. Originally, the term *RNA interference* (RNAi) referred to the actions

of the related class of regulatory noncoding RNAs called *small interfering RNAs* (siRNAs); RNAi now encompasses the related microRNAs as well.

16. Schickel, Boyerinas, et al. (2008).

17. Georgantas, Hildreth, et al. (2007).

18. See, for example, Stocum (2004), and Straube and Tanaka (2006).

19. For dedifferentiation in repair of cartilage, see Schulze-Tanzil (2009). For dedifferentiation in repair of the peripheral nervous system, see Chen, Yu, et al. (2007). Bonventre (2003) discusses dedifferentiation in kidney repair.

20. Stocum (2002). Interestingly, biochemicals obtained from amphibians can boost regeneration in mammals, an indication that the mammalian genome can epigenetically respond in an ordered way to environmental influences to which it is never normally exposed.

21. Fibroblast cells were used in these experiments (see Takahashi, Okita, et al. 2007); see also Diez-Torre, Andrade, et al. (2009), and Lyssiotis, Foreman, et al. (2009). Kim, Zaehres, et al. (2008) used neural stem cells to generate pluripotent cells. The pluripotent cells generated in these experiments are called induced pluripotent stem cells (iPSCs) to distinguish them from actual embryonic stem cells (ESCs); iPSCs seem to have the essential properties of ESCs, including the capacity to differentiate into all three of the primary germ layers, but they may have subtle differences (Ou, Wang, et al. 2010). Araki, Jincho, et al. (2010) document the stages of dedifferentiation from adult fibroblast cell to iPSC.

22. Okano (2009). Fibroblasts can also be transformed into neurons without going through a pluripotent stage (Masip, Veiga, et al. 2010). This process is known as transdifferentiation (Collas and Hakelien 2003).

23. This line of research stems from an important study by Mintz and Illmensee (1975), which demonstrated that when malignant mouse teratocarcinoma cells are transplanted into the blastocyst (the mammalian form of blastula) of a mouse, they are normalized and contribute to the formation of a variety of normal cell types. Recently, Hochedlinger, Blelloch, et al. (2004) transplanted the genome of a mouse melanoma cell into an enucleated oocyte, from which they derived normal embryonic stem cells. From these stem cells, they generated normal-appearing mice.

24. Kulesa, Kasemeier-Kulesa, et al. (2006); Hendrix, Seftor, et al. (2007).

25. Collas (2010) is typical in this regard, in the context of cellular differentiation.

26. For the minimalist notion of programming in situated robotics, see Hendriks-Jansen (1996). Wolfram (2002) is an expanded, semi-mystical view of the significance of a minimalist program, motivated by Wolfram's research on cellular automata.

27. Passier and Mummery (2003).

28. Moreover, there are subtle differences between embryonic stem cells (ESCs) and induced embryonic stem cells (iPSCs) that are clinically relevant. For example, ESCs have proved much more efficient in promoting neuronal redifferentiation than iPSCs have (Tokumoto, Ogawa, et al. 2010).

29. This is how Conrad Waddington, who coined the term *epigenetic*, describes its derivation: "Some years ago I introduced the word 'epigenetic,' derived from the Aristotelian word 'epigenesis,' which had more or less passed into disuse, as a suitable name for the branch of biology which studies the causal interactions between genes and their products which bring the phenotype into being" (Waddington 1968).

30. In essence, Waddington's goal was a synthesis of what was then known as "embryology" and genetics; we now call this synthesis developmental biology. I consider Waddington the father of modern developmental biology.

31. With respect to the causal primacy of genome or cytoplasm within a cell, Waddington said, "Of course, to insist on pursuing the argument *ad infinitum* leads to a ridiculous question, like asking whether the hen or the egg came first, because finally the gene and the cytoplasm are dependent on each other and neither could exist alone" (Waddington 1935/1946). This is a good encapsulation of the executive cell view.

32. The responsiveness of the genome is conveyed in the following quote from Waddington (1962): ". . . the almost universal occurrence in higher organisms of feedback between cytoplasm and genes, such that the nature of the cytoplasm determines the intensity of the syntheses controlled by the various genes in the nucleus." This description is also a nice characterization of modern epigenetics.

33. See, for example, Gurdon and Melton (2008), and de Souza (2010).

Chapter 11. **Pray for the Devil**

1. This was Hobbes's assessment—in his magnum opus, *Leviathan*—of the human condition in the state of nature, that is, prior to the civilizing

influence of the state. In my edition (edited by R. Tuck), the actual quote is "And the life of man, solitary, poore, nasty, brutish and short." (Hobbes, 1651/1996).

2. Wroe, McHenry, et al. (2005).

3. McCallum (2008).

4. This mechanism by which cancer cells are directly transmitted from one individual to another is called *allografting* (Pearse and Swift 2006).

5. The poor immune recognition has been attributed to low diversity at major histocompatibility complex (MHC) loci, the protein products of which, in certain immune cells, are responsible for recognizing the difference between self and nonself (Siddle, Kreiss, et al. 2007). Low genetic variation at these sites makes a seeming match between self and nonself more likely. Murgia, Pritchard, et al. (2006) have proposed that the reason for high MHC diversity typically found in most animals is to prevent contagious cancers.

6. For the bottleneck in cheetah populations, see O'Brien, Wildt, et al. (1983). For tolerance of skin allografts, see Sanjayan and Crooks (1996).

7. Murgia, Pritchard, et al. (2006).

8. Hsiao, Liao, et al. (2008).

9. Pearse and Swift (2006).

10. See, for example, Daley (2008).

11. Loh, Hayes, et al. (2006). Murchison, Tovar, et al. (2010), however, proposed that DFTD cells are derived from Schwann cells, a type of glia (support cells in the nervous system) that supply nutrients to axons in the peripheral nervous system.

12. Tu, Lin, et al. (2002); Sales, Winslet, et al. (2007); Trosko (2009).

13. Schulz and Hatina (2006).

14. Johnsen, Malene Krag, et al. (2009).

15. Curtis (1965); Frank and Nowak (2004). Gatenby and Vincent (2003) is a good summarization of SMT.

16. Hisamuddin and Yang (2006).

17. Duesberg (2005); Duesberg, Li, et al. (2005); Nicholson and Duesberg (2009).

18. Bharadwaj and Yu (2004); Pathak and Multani (2006).

19. Duesberg, Li, et al. (2000); Pezer and Ugarkovic (2008).

20. DFTD is a clonal cell line that originated in a single individual some time prior to 1996.

21. CTVT originated between two hundred and fifty and twenty-five hundred years ago (Murgia, Pritchard, et al. 2006). Frank (2007) considers CTVT a distinct genomic species, the "malignant dog."

22. Feinberg, Ohlsson, et al. (2006); Suijkerbuijk, van der Wall, et al. (2007).

23. Feinberg, Ohlsson, et al. (2006).

24. Gaudet, Hodgson, et al. (2003).

25. Feinberg, Ohlsson, et al. (2006).

26. Lotem and Sachs (2002).

27. Feinberg, Ohlsson, et al. (2006).

28. See, for example, Ganesan, Nolan, et al. (2009).

29. Fassati and Mitchison (2010).

30. Capp (2005) is a good introduction to this approach. See also Ingber (2002), Soto and Sonnenschein (2004), and Chung, Baseman, et al. (2005).

31. Kenny and Bissell (2003). See also Bissell and Labarge (2005), Nelson and Bissell (2006), and Kenny, Lee, et al. (2007).

Bibliography

Aagaard-Tillery, K. M., K. Grove, et al. (2008). "Developmental origins of disease and determinants of chromatin structure: Maternal diet modifies the primate fetal epigenome." *J Mol Endocrinol* **41**(2): 91–102.

Allen, G. (1978). *Thomas Hunt Morgan: The man and his science.* Princeton: Princeton University Press.

Amor, D. J., and J. Halliday (2008). "A review of known imprinting syndromes and their association with assisted reproduction technologies." *Hum Reprod* **23**(12): 2826–2834.

Anway, M. D., A. S. Cupp, et al. (2005). "Epigenetic transgenerational actions of endocrine disruptors and male fertility." *Science* **308**(5727): 1466–1469.

Anway, M. D., and M. K. Skinner (2008). "Epigenetic programming of the germ line: Effects of endocrine disruptors on the development of transgenerational disease." *Reprod Biomed Online* **16**(1): 23–25.

Araki, R., Y. Jincho, et al. (2010). "Conversion of ancestral fibroblasts to induced pluripotent stem cells." *Stem Cells* **28**(2): 213–220.

Ashabranner, B. K. (1988). *Always to remember: The story of the Vietnam Veterans Memorial.* New York: Putnam.

Atlan, H., and M. Koppel (1990). "The cellular computer DNA: Program or data." *Bull Math Biol* **52**(3): 335–348.

Au, T. M., A. K. Greenwood, et al. (2006). "Differential social regulation of two pituitary gonadotropin-releasing hormone receptors." *Behav Brain Res* **170**(2): 342–346.

Bailey, J. A., K. G. Hill, et al. (2009). "Parenting practices and problem behavior across three generations: Monitoring, harsh discipline, and drug use in the intergenerational transmission of externalizing behavior." *Dev Psychol* **45**(5): 1214–1226.

Ballestar, E., M. Esteller, et al. (2006). "The epigenetic face of systemic lupus erythematosus." *J Immunol* **176**(12): 7143–7147.

Bardi, M., and M. A. Huffman (2002). "Effects of maternal style on infant behavior in Japanese macaques (*Macaca fuscata*)." *Dev Psychobiol* **41**(4): 364–372.

Bardi, M., and M. A. Huffman (2006). "Maternal behavior and maternal stress are associated with infant behavioral development in macaques." *Dev Psychobiol* **48**(1): 1–9.

Barker, M., S. Robinson, et al. (1997). "Birth weight and body fat distribution in adolescent girls." *Arch Dis Child* **77**(5): 381–383.

Bartholdi, D., M. Krajewska-Walasek, et al. (2009). "Epigenetic mutations of the imprinted IGF2-H19 domain in Silver-Russell syndrome (SRS): Results from a large cohort of patients with SRS and SRS-like phenotypes." *J Med Genet* **46**(3): 192–197.

Beadle, G. W., and E. L. Tatum (1941). "Genetic control of biochemical reactions in *Neurospora*." *Proc Natl Acad Sci USA* **27**(11): 499–506.

Beck, B., and M. Power (1988). "Correlates of sexual and maternal competence in captive gorillas." *Zoo Biol* **7**: 339–350.

Beck, S., A. Olek, et al. (1999). "From genomics to epigenomics: A loftier view of life." *Nat Biotechnol* **17**(12): 1144.

Bellinger, L., and S. C. Langley-Evans (2005). "Fetal programming of appetite by exposure to a maternal low-protein diet in the rat." *Clin Sci (Lond)* **109**(4): 413–420.

Belshaw, R., V. Pereira, et al. (2004). "Long-term reinfection of the human genome by endogenous retroviruses." *Proc Natl Acad Sci USA* **101**(14): 4894–4899.

Berletch, J. B., F. Yang, et al. (2010). "Escape from X inactivation in mice and humans." *Genome Biol* **11**(6): 213.

Bernal, A. J., and R. L. Jirtle (2010). "Epigenomic disruption: The effects of early developmental exposures." *Birth Defects Res A Clin Mol Teratol* **88**(10): 938–944.

Bettegowda, A., K. Lee, et al. (2007). "Cytoplasmic and nuclear determinants of the maternal-to-embryonic transition." *Reprod Fertil Dev* **20**(1): 45–53.

Beurton, P. J., R. Falk, et al. (2000). *The concept of the gene in development and evolution: Historical and epistemological perspectives.* Cambridge, UK: Cambridge University Press.

Bevilacqua, E., R. Brunelli, et al. (2010). "Review and meta-analysis: Benefits and risks of multiple courses of antenatal corticosteroids." *J Matern Fetal Neonatal Med* **23**(4): 244–260.

Bharadwaj, R., and H. Yu (2004). "The spindle checkpoint, aneuploidy, and cancer." *Oncogene* **23**(11): 2016–2027.

Bianco, S. D., and U. B. Kaiser (2009). "The genetic and molecular basis of idiopathic hypogonadotropic hypogonadism." *Nat Rev Endocrinol* **5**(10): 569–576.

Biliya, S., and L. A. Bulla, Jr. (2010). "Genomic imprinting: The influence of differential methylation in the two sexes." *Exp Biol Med (Maywood)* **235**(2): 139–147.

Bissell, M. J., and M. A. Labarge (2005). "Context, tissue plasticity, and cancer: Are tumor stem cells also regulated by the microenvironment?" *Cancer Cell* **7**(1): 17–23.

Bittel, D. C., N. Kibiryeva, et al. (2005). "Microarray analysis of gene/transcript expression in Angelman syndrome: Deletion versus UPD." *Genomics* **85**(1): 85–91.

Bjorntorp, P., and R. Rosmond (2000). "Obesity and cortisol." *Nutrition* **16**(10): 924–936.

Blewitt, M. E., N. K. Vickaryous, et al. (2006). "Dynamic reprogramming of DNA methylation at an epigenetically sensitive allele in mice." *PLoS Genet* **2**(4): e49.

Blobel, G. (1980). "Intracellular protein topogenesis." *Proc Natl Acad Sci USA* **77**(3): 1496–1500.

Bodemer, C. W. (1964). "Regeneration and the decline of preformationism in eighteenth century embryology." *Bull Hist Med* **38**: 20–31.

Bonventre, J. V. (2003). "Dedifferentiation and proliferation of surviving epithelial cells in acute renal failure." *J Am Soc Nephrol* **14**(Suppl 1): S55–S61.

Booth, A., G. Shelley, et al. (1989). "Testosterone, and winning and losing in human competition." *Horm Behav* **23**(4): 556–571.

Bradley, R. G., E. B. Binder, et al. (2008). "Influence of child abuse on adult depression: Moderation by the corticotropin-releasing hormone receptor gene." *Arch Gen Psychiatry* **65**(2): 190–200.

Brand, S. R., S. M. Engel, et al. (2006). "The effect of maternal PTSD following in utero trauma exposure on behavior and temperament in the 9-month-old infant." *Ann N Y Acad Sci* **1071**: 454–458.

Brenner, S., G. Elgar, et al. (1993). "Characterization of the pufferfish (*Fugu*) genome as a compact model vertebrate genome." *Nature* **366**(6452): 265–268.

Brown, A. S., J. van Os, et al. (2000). "Further evidence of relation between prenatal famine and major affective disorder." *Am J Psychiatry* **157**(2): 190–195.

Brown, C., and J. Greally (2003). "A stain upon the silence: Genes escaping X inactivation." *Trends Genet* **19**: 432–438.

Burdge, G. C., M. A. Hanson, et al. (2007). "Epigenetic regulation of transcription: A mechanism for inducing variations in phenotype (fetal programming) by differences in nutrition during early life?" *Br J Nutr* **97**(6): 1036–1046.

Burmeister, S. S., V. Kailasanath, et al. (2007). "Social dominance regulates androgen and estrogen receptor gene expression." *Horm Behav* **51**(1): 164–170.

Calatayud, F., and C. Belzung (2001). "Emotional reactivity in mice, a case of nongenetic heredity?" *Physiol Behav* **74**(3): 355–362.

Calin, G. A., C. D. Dumitru, et al. (2002). "Frequent deletions and downregulation of micro-RNA genes *miR15* and *miR16* at 13q14 in chronic lymphocytic leukemia." *Proc Natl Acad Sci USA* **99**(24): 15524–15529.

Callinan, P. A., and A. P. Feinberg (2006). "The emerging science of epigenomics." *Hum Mol Genet* **15**(Suppl 1): R95–R101.

Campbell, K. H., J. McWhir, et al. (1996). "Sheep cloned by nuclear transfer from a cultured cell line." *Nature* **380**(6569): 64–66.

Capp, J. P. (2005). "Stochastic gene expression, disruption of tissue averaging effects, and cancer as a disease of development." *BioEssays* **27**(12): 1277–1285.

Carninci, P., and Y. Hayashizaki (2007). "Noncoding RNA transcription beyond annotated genes." *Curr Opin Genet Dev* **17**(2): 139–144.

Carrell, D. T., and S. S. Hammoud (2010). "The human sperm epigenome and its potential role in embryonic development." *Mol Hum Reprod* **16**(1): 37–47.

Cassidy, S. B., and D. H. Ledbetter (1989). "Prader-Willi syndrome." *Neurol Clin* **7**(1): 37–54.

Castle, W. E., F. W. Carpenter, et al. (1906). "The effects of inbreeding, cross-breeding, and selection upon fertility and variability of *Drosophila*." *Proc Am Acad Arts Sci* **41**.

Castle, W. E., and S. Wright (1916). "Studies of inheritance in guinea pigs and rats." *Carnegie Inst Wash Publ* **241**: 163–190.

Cattanach, B. M., C. V. Beechey, et al. (2006). "Interactions between imprinting effects: Summary and review." *Cytogenet Genome Res* **113**(1–4): 17–23.

Champagne, F. A., and J. P. Curley (2009). "Epigenetic mechanisms mediating the long-term effects of maternal care on development." *Neurosci Biobehav Rev* **33**(4): 593–600.

Champagne, F., J. Diorio, et al. (2001). "Naturally occurring variations in maternal behavior in the rat are associated with differences in estrogen-inducible central oxytocin receptors." *Proc Natl Acad Sci USA* **98**(22): 12736–12741.

Champagne, F. A, and M. J. Meaney (2001). "Like mother, like daughter: Evidence for non-genomic transmission of parental behavior and stress responsivity." *Prog Brain Res* **133**: 287–302.

Champagne, F. A., I. C. Weaver, et al. (2006). "Maternal care associated with methylation of the estrogen receptor-alpha1b promoter and estrogen receptor-alpha expression in the medial preoptic area of female offspring." *Endocrinology* **147**(6): 2909–2915.

Champoux, M., E. Byrne, et al. (1992). "Motherless mothers revisited: Rhesus maternal behavior and rearing history." *Primates* **33**: 251–255.

Chang, H. S., M. D. Anway, et al. (2006). "Transgenerational epigenetic imprinting of the male germline by endocrine disruptor exposure during gonadal sex determination." *Endocrinology* **147**(12): 5524–5541.

Chen, C., J. Visootsak, et al. (2007). "Prader-Willi syndrome: An update and review for the primary pediatrician." *Clin Pediatr (Phila)* **46**(7): 580–591.

Chen, Z. L., W. M. Yu, et al. (2007). "Peripheral regeneration." *Annu Rev Neurosci* **30**: 209–233.

Chong, S., N. A. Youngson, et al. (2007). "Heritable germline epimutation is not the same as transgenerational epigenetic inheritance." *Nat Genet* **39**(5): 574–575, author reply 575–576.

Chow, J. C., Z. Yen, et al. (2005). "Silencing of the mammalian X chromosome." *Annu Rev Genomics Hum Genet* **6**: 69–92.

Christensen, B. C., E. A. Houseman, et al. (2009). "Aging and environmental exposures alter tissue-specific DNA methylation dependent upon CpG island context." *PLoS Genet* **5**(8): e1000602.

Christian, J. C., D. Bixler, et al. (1971). "Hypogandotropic hypogonadism with anosmia: The Kallmann syndrome." *Birth Defects Orig Artic Ser* **7**(6): 166–171.

Chung, L. W., A. Baseman, et al. (2005). "Molecular insights into prostate cancer progression: The missing link of tumor microenvironment." *J Urol* **173**(1): 10–20.

Chung, Y., C. E. Bishop, et al. (2009). "Reprogramming of human somatic cells using human and animal oocytes." *Cloning Stem Cells* **11**(2): 213–223.

Cohn, S. A., D. S. Emmerich, et al. (1989). "Differences in the responses of heterozygous carriers of colorblindness and normal controls to briefly presented stimuli." *Vision Res* **29**(2): 255–262.

Collas, P. (2010). "Programming differentiation potential in mesenchymal stem cells." *Epigenetics* **5**(6).

Collas, P., and A. M. Hakelien (2003). "Teaching cells new tricks." *Trends Biotechnol* **21**(8): 354–361.

Cooper, W. N., R. Curley, et al. (2007). "Mitotic recombination and uniparental disomy in Beckwith-Wiedemann syndrome." *Genomics* **89**(5): 613–617.

Costa, F. F. (2008). "Non-coding RNAs, epigenetics and complexity." *Gene* **410**(1): 9–17.

Coventry, W. L., S. E. Medland, et al. (2009). "Phenotypic and discordant-monozygotic analyses of stress and perceived social support as antecedents to or sequelae of risk for depression." *Twin Res Hum Genet* **12**(5): 469–488.

Crespi, B. (2008). "Genomic imprinting in the development and evolution of psychotic spectrum conditions." *Biol Rev Camb Philos Soc* **83**(4): 441–493.

Crews, D. (2010). "Epigenetics, brain, behavior, and the environment." *Hormones (Athens)* **9**(1): 41–50.

Cropley, J. E., C. M. Suter, et al. (2006). "Germ-line epigenetic modification of the murine *Avy* allele by nutritional supplementation." *Proc Natl Acad Sci USA* **103**(46): 17308–17312.

Currenti, S. A. (2010). "Understanding and determining the etiology of autism." *Cell Mol Neurobiol* **30**(2): 161–171.

Curtis, H. J. (1965). "Formal discussion of: Somatic mutations and carcinogenesis." *Cancer Res* **25**: 1305–1308.

Dai, Y., L. Wang, et al. (2006). "Fate of centrosomes following somatic cell nuclear transfer (SCNT) in bovine oocytes." *Reproduction* **131**(6): 1051–1061.

Daley, G. Q. (2008). "Common themes of dedifferentiation in somatic cell reprogramming and cancer." *Cold Spring Harb Symp Quant Biol* **73**: 171–174.

Damcott, C. M., P. Sack, et al. (2003). "The genetics of obesity." *Endocrinol Metab Clin North Am* **32**(4): 761–786.

Deakin, J. E., J. Chaumeil, et al. (2009). "Unravelling the evolutionary origins of X chromosome inactivation in mammals: Insights from marsupials and monotremes." *Chromosome Res* **17**(5): 671–685.

Dean, F., C. Yu, et al. (2001). "Prenatal glucocorticoid modifies hypothalamo-pituitary-adrenal regulation in prepubertal guinea pigs." *Neuroendocrinology* **73**(3): 194–202.

De Boo, H. A., and J. E. Harding (2006). "The developmental origins of adult disease (Barker) hypothesis." *Aust N Z J Obstet Gynaecol* **46**(1): 4–14.

Deeb, S. S. (2005). "The molecular basis of variation in human color vision." *Clin Genet* **67**(5): 369–377.

Delage, B., and R. H. Dashwood (2008). "Dietary manipulation of histone structure and function." *Annu Rev Nutr* **28**: 347–366.

Delaval, K., A. Wagschal, et al. (2006). "Epigenetic deregulation of imprinting in congenital diseases of aberrant growth." *BioEssays* **28**(5): 453–459.

Denenberg, V. H., and K. M. Rosenberg (1967). "Nongenetic Transmission of Information." *Nature* **216**(5115): 549–550.

de Souza, N. (2010). "Primer: Induced pluripotency." *Nat Methods* **7**(1): 20–21.

Diamanti-Kandarakis, E., J. P. Bourguignon, et al. (2009). "Endocrine-disrupting chemicals: An Endocrine Society scientific statement." *Endocr Rev* **30**(4): 293–342.

DiBartolo, P. M., and M. Helt (2007). "Theoretical models of affectionate versus affectionless control in anxious families: A critical examination based on observations of parent-child interactions." *Clin Child Fam Psychol Rev* **10**(3): 253–274.

Diez-Torre, A., R. Andrade, et al. (2009). "Reprogramming of melanoma cells by embryonic microenvironments." *Int J Dev Biol* **53**(8–10): 1563–1568.

Dobyns, W. B., A. Filauro, et al. (2004). "Inheritance of most X-linked traits is not dominant or recessive, just X-linked." *Am J Med Genet A* **129A**(2): 136–143.

Dolinoy, D. C., D. Huang, et al. (2007). "Maternal nutrient supplementation counteracts bisphenol A–induced DNA hypomethylation in early development." *Proc Natl Acad Sci USA* **104**(32): 13056–13061.

Dolinoy, D. C., J. R. Weidman, et al. (2006). "Maternal genistein alters coat color and protects Avy mouse offspring from obesity by modifying the fetal epigenome." *Environ Health Perspect* **114**(4): 567–572.

Drake, A. J., J. I. Tang, et al. (2007). "Mechanisms underlying the role of glucocorticoids in the early life programming of adult disease." *Clin Sci (Lond)* **113**(5): 219–232.

Driscoll, D. J., M. F. Waters, et al. (1992). "A DNA methylation imprint, determined by the sex of the parent, distinguishes the Angelman and Prader-Willi syndromes." *Genomics* **13**(4): 917–924.

Duesberg, P. (2005). "Does aneuploidy or mutation start cancer?" *Science* **307**(5706): 41.

Duesberg, P., R. Li, et al. (2000). "Aneuploidy precedes and segregates with chemical carcinogenesis." *Cancer Genet Cytogenet* **119**(2): 83–93.

Duesberg, P., R. Li, et al. (2005). "The chromosomal basis of cancer." *Cell Oncol* **27**(5–6): 293–318.

Duhl, D. M., M. E. Stevens, et al. (1994). "Pleiotropic effects of the mouse lethal yellow (Ay) mutation explained by deletion of a maternally expressed gene and the simultaneous production of agouti fusion RNAs." *Development* **120**(6): 1695–1708.

Duhl, D. M., H. Vrieling, et al. (1994). "Neomorphic agouti mutations in obese yellow mice." *Nat Genet* **8**(1): 59–65.

Eddy, S. R. (2001). "Non-coding RNA genes and the modern RNA world." *Nat Rev Genet* **2**(12): 919–929.

Eilertsen, K. J., R. A. Power, et al. (2007). "Targeting cellular memory to reprogram the epigenome, restore potential, and improve somatic cell nuclear transfer." *Anim Reprod Sci* **98**(1–2): 129–146.

Elgar, G., and T. Vavouri (2008). "Tuning in to the signals: Noncoding sequence conservation in vertebrate genomes." *Trends Genet* **24**(7): 344–352.

Emack, J., A. Kostaki, et al. (2008). "Chronic maternal stress affects growth, behaviour and hypothalamo-pituitary-adrenal function in juvenile offspring." *Horm Behav* **54**(4): 514–520.

Emslie, C., and K. Hunt (2008). "The weaker sex? Exploring lay understandings of gender differences in life expectancy: A qualitative study." *Soc Sci Med* **67**(5): 808–816.

Engert, V., R. Joober, et al. (2009). "Behavioral response to methylphenidate challenge: Influence of early life parental care." *Dev Psychobiol* **51**(5): 408–416.

Erwin, J. A., and J. T. Lee (2008). "New twists in X-chromosome inactivation." *Curr Opin Cell Biol* **20**(3): 349–355.

Evanson, N. K., J. G. Tasker, et al. (2010). "Fast feedback inhibition of the HPA axis by glucocorticoids is mediated by endocannabinoid signaling." *Endocrinology* **151**(10) 4811–4819.

Fassati, A., and N. A. Mitchison (2010). "Testing the theory of immune selection in cancers that break the rules of transplantation." *Cancer Immunol Immunother* **59**(5): 643–651.

Feinberg, A. P., R. Ohlsson, et al. (2006). "The epigenetic progenitor origin of human cancer." *Nat Rev Genet* **7**(1): 21–33.

Fishman, L., and J. H. Willis (2006). "A cytonuclear incompatibility causes anther sterility in *Mimulus* hybrids." *Evolution* **60**(7): 1372–1381.

Forterre, P. (2001). "Genomics and early cellular evolution. The origin of the DNA world." *C R Acad Sci Ser III* **324**(12): 1067–1076.

Forterre, P. (2002). "The origin of DNA genomes and DNA replication proteins." *Curr Opin Microbiol* **5**(5): 525–532.

Fowden, A. L., C. Sibley, et al. (2006). "Imprinted genes, placental development and fetal growth." *Horm Res* **65**(Suppl 3): 50–58.

Fox Keller, E. (1999). "Elusive locus of control in biological development: Genetic versus developmental programs." *Exp Zool* **285**(3): 283–290.

Fox Keller, E. (2000). *The century of the gene.* Cambridge: Harvard University Press.

Francis, D. D., F. A. Champagne, et al. (1999). "Maternal care, gene expression, and the development of individual differences in stress reactivity." *Ann N Y Acad Sci* **896**: 66–84.

Francis, D. D., F. C. Champagne, et al. (2000). "Variations in maternal behaviour are associated with differences in oxytocin receptor levels in the rat." *J Neuroendocrinol* **12**(12): 1145–1148.

Francis, D., J. Diorio, et al. (1999). "Nongenomic transmission across generations of maternal behavior and stress responses in the rat." *Science* **286**(5442): 1155–1158.

Francis, D. D., and M. J. Meaney (1999). "Maternal care and the development of stress responses." *Curr Opin Neurobiol* **9**(1): 128–134.

Francis, R. C. (1992). "Sexual lability in teleosts: Developmental factors." *Q Rev Biol* **67**(1): 1–18.

Francis, R. C., B. Jacobson, et al. (1992). "Hypertrophy of gonadotropin releasing hormone-containing neurons after castration in the teleost, *Haplochromis burtoni.*" *J Neurobiol* **23**(8): 1084–1093.

Francis, R. C., K. Soma, et al. (1993). "Social regulation of the brain-pituitary-gonadal axis." *Proc Natl Acad Sci USA* **90**: 7794–7798.

Frank, S. A., and M. A. Nowak (2004). "Problems of somatic mutation and cancer." *BioEssays* **26**(3): 291–299.

Frank, U. (2007). "The evolution of a malignant dog." *Evol Dev* **9**(6): 521–522.

French, M., M. Venu, et al. (2009). "Non-identical Kallmann's syndrome in otherwise identical twins." *Endocr Abstr* **19**: 46.

Fulka, J., Jr., and H. Fulka (2007). "Somatic cell nuclear transfer (SCNT) in mammals: The cytoplast and its reprogramming activities." *Adv Exp Med Biol* **591**: 93–102.

Galvan, A., F. S. Falvella, et al. (2010). "Genome-wide association study in discordant sibships identifies multiple inherited susceptibility alleles linked to lung cancer." *Carcinogenesis* **31**(3): 462–465.

Ganesan, A., L. Nolan, et al. (2009). "Epigenetic therapy: Histone acetylation, DNA methylation and anti-cancer drug discovery." *Curr Cancer Drug Targets* **9**(8): 963–981.

Gatenby, R. A., and T. L. Vincent (2003). "An evolutionary model of carcinogenesis." *Cancer Res* **63**(19): 6212–6220.

Gaudet, F., J. G. Hodgson, et al. (2003). "Induction of tumors in mice by genomic hypomethylation." *Science* **300**(5618): 489–492.

Georgantas, R. W., III, R. Hildreth, et al. (2007). "CD34+ hematopoietic stem-progenitor cell microRNA expression and function: A circuit diagram of differentiation control." *Proc Natl Acad Sci USA* **104**(8): 2750–2755.

Gibson, G. (2007). "Human evolution: thrifty genes and the Dairy Queen." *Curr Biol* **17**(8): R295–R296.

Gilbert, S. F. (1991). *Developmental biology*, 3rd ed. Sunderland, MA: Sinauer.

Gilbert, S. F., and S. Sarkar (2000). "Embracing complexity: Organicism for the 21st century." *Dev Dyn* **219**(1): 1–9.

Goldberg, J., W. R. True, et al. (1990). "A twin study of the effects of the Vietnam War on posttraumatic stress disorder." *JAMA* **263**(9): 1227–1232.

Grace, K. S., and K. D. Sinclair (2009). "Assisted reproductive technology, epigenetics, and long-term health: A developmental time bomb still ticking." *Semin Reprod Med* **27**(5): 409–416.

Greenfield, E. A., and N. F. Marks (2010). "Identifying experiences of physical and psychological violence in childhood that jeopardize mental health in adulthood." *Child Abuse Negl* **34**(3): 161–171.

Griffiths, P., and R. D. Gray (1994). "Developmental systems and evolutionary explanation." *J Phil* **91**: 277–304.

Griffiths, P., and E. Neumann-Held (1999). "The many faces of the gene." *BioScience* **49**(8): 656–662.

Gross, K. L., and J. A. Cidlowski (2008). "Tissue-specific glucocorticoid action: A family affair." *Trends Endocrinol Metab* **19**(9): 331–339.

Gross-Sorokin, M. Y., S. D. Roast, et al. (2006). "Assessment of feminization of male fish in English rivers by the Environment Agency of England and Wales." *Environ Health Perspect* **114**(Suppl 1): 147–151.

Guerriero, G. (2009). "Vertebrate sex steroid receptors: Evolution, ligands, and neurodistribution." *Ann N Y Acad Sci* **1163**: 154–168.

Gurdon, J. B., and D. A. Melton (2008). "Nuclear reprogramming in cells." *Science* **322**(5909): 1811–1815.

Hales, C. N., and D. J. P. Barker (2001). "The thrifty phenotype hypothesis: Type 2 diabetes." *Br Med Bull* **60**(1): 5–20.

Hamelin, C. E., G. Anglin, et al. (2006). "Genomic imprinting in Turner syndrome: Effects on response to growth hormone and on risk of sensorineural hearing loss." *J Clin Endocrinol Metab* **91**(8): 3002–3010.

Hannes, R.-P., D. Franck, et al. (1984). "Effects of rank-order fights on whole-body and blood concentrations of androgens and corticosteroids in the male swordtail (*Xiphophorus helleri*)." *Z Tierpsychol* **65**: 53–65.

Haque, F. N., Gottesman, I. I., et al. (2009). "Not really identical: Epigenetic differences in monozygotic twins and implications for twin studies in psychiatry." *Am J Med Genet C Semin Med Genet* **151C**(2): 136–141.

Harlow, H. F., M. K. Harlow, et al. (1971). "From thought to therapy: Lessons from a primate laboratory." *Am Sci* **50**: 538–549.

Harlow, H. F., and R. R. Zimmerman (1959). "Affectional responses in the infant monkey." *Science* **136**: 421–431.

Hatchwell, E., and J. M. Greally (2007). "The potential role of epigenomic dysregulation in complex human disease." *Trends Genet* **23**(11): 588–595.

Hayashi, T., A. G. Motulsky, et al. (1999). "Position of a 'green-red' hybrid gene in the visual pigment array determines colour-vision phenotype." *Nat Genet* **22**(1): 90–93.

Hayes, T. B., A. A. Stuart, et al. (2006). "Characterization of atrazine-induced gonadal malformations in African clawed frogs (*Xenopus laevis*) and comparisons with effects of an androgen antagonist (cyproterone acetate) and exogenous estrogen (17β-estradiol): Support for the demasculinization/feminization hypothesis." *Environ Health Perspect* **114**(Suppl 1): 134–141.

Henderson, I. R., and S. E. Jacobsen (2007). "Epigenetic inheritance in plants." *Nature* **447**(7143): 418–424.

Hendriks-Jansen, H. (1996). *Catching ourselves in the act: Situated activity, integrative emergence, evolution, and human thought*. Cambridge, MA: MIT Press.

Hendrix, M. J., E. A. Seftor, et al. (2007). "Reprogramming metastatic tumour cells with embryonic microenvironments." *Nat Rev Cancer* **7**(4): 246–255.

Hess, C. T. (2009). "Monitoring laboratory values: Vitamin B1, vitamin B6, vitamin B12, folate, calcium, and magnesium." *Adv Skin Wound Care* **22**(6): 288.

Hinney, A., C. I. Vogel, et al. (2010). "From monogenic to polygenic obesity: Recent advances." *Eur Child Adolesc Psychiatry* **19**(3): 297–310.

Hipkin, L. J., I. F. Casson, et al. (1990). "Identical twins discordant for Kallmann's syndrome." *J Med Genet* **27**: 198–199.

Hipp, J., and A. Atala (2008). "Sources of stem cells for regenerative medicine." *Stem Cell Rev* **4**(1): 3–11.

Hisamuddin, I. M., and V. W. Yang (2006). "Molecular genetics of colorectal cancer: An overview." *Curr Colorectal Cancer Rep* **2**(2): 53–59.

Hobbes, T. (1651/1996). *Leviathan*, ed. R. Tuck. New York: Cambridge University Press.

Hoch, S. L. (1998). "Famine, disease, and mortality patterns in the parish of Borshevka, Russia, 1830–1912." *Popul Stud (Camb)* **52**(3): 357–368.

Hochedlinger, K., R. Blelloch, et al. (2004). "Reprogramming of a melanoma genome by nuclear transplantation." *Genes Dev* **18**(15): 1875–1885.

Holliday, R. (1996). "Endless quest." *BioEssays* **18**(1): 3–5.

Holliday, R. (2006). "Epigenetics: A historical overview." *Epigenetics* **1**(2): 76–80.

Horvitz, H. R., and J. E. Sulston (1980). "Isolation and genetic characterization of cell-lineage mutants of the nematode *Caenorhabditis elegans*." *Genetics* **96**(2): 435–454.

Hsiao, Y. W., K. W. Liao, et al. (2008). "Interactions of host IL-6 and IFN-gamma and cancer-derived TGF-beta1 on MHC molecule expression during tumor spontaneous regression." *Cancer Immunol Immunother* **57**(7): 1091–1104.

Hunt, D. M., A. J. Williams, et al. (1993). "Structure and evolution of the polymorphic photopigment gene of the marmoset." *Vision Res* **33**(2): 147–154.

Ikeda, D., and S. Watabe (2004). "[Fugu genome: The smallest genome size in vertebrates]." *Tanpakushitsu Kakusan Koso* **49**(14): 2235–2241.

Ingber, D. E. (2002). "Cancer as a disease of epithelial-mesenchymal interactions and extracellular matrix regulation." *Differentiation* **70**(9–10): 547–560.

Jablonka, E. (2004). "The evolution of the peculiarities of mammalian sex chromosomes: An epigenetic view." *BioEssays* **26**: 1327–1332.

Jablonka, E., and M. J. Lamb (2002). "The changing concept of epigenetics." *Ann N Y Acad Sci* **981**: 82–96.

Jablonka, E., and G. Raz (2009). "Transgenerational epigenetic inheritance: Prevalence, mechanisms, and implications for the study of heredity and evolution." *Q Rev Biol* **84**(2): 131–176.

Jacobs, G. H. (1998). "A perspective on color vision in platyrrhine monkeys." *Vision Res* **38**(21): 3307–3313.

Jacobs, G. H. (2008). "Primate color vision: a comparative perspective." *Vis Neurosci* **25**(5–6): 619–633.

Jacobs, G. H., and J. F. Deegan II (2003). "Cone pigment variations in four genera of New World monkeys." *Vision Res* **43**(3): 227–236.

Jameson, K. A., S. M. Highnote, et al. (2001). "Richer color experience in observers with multiple photopigment opsin genes." *Psychon Bull Rev* **8**(2): 244–261.

Jobling, S., R. Williams, et al. (2006). "Predicted exposures to steroid estrogens in U.K. rivers correlate with widespread sexual disruption in wild fish populations." *Environ Health Perspect* **114**(Suppl 1): 32–39.

Johnsen, H., K. Malene Krag, et al. (2009). "Cancer stem cells and the cellular hierarchy in haematological malignancies." *Eur J Cancer* **45**: 194–201.

Jones, J. R., C. Skinner, et al. (2008). "Hypothesis: Dysregulation of methylation of brain-expressed genes on the X chromosome and autism spectrum disorders." *Am J Med Genet A* **146A**(17): 2213–2220.

Jones, P. A., and S. B. Baylin (2007). "The epigenomics of cancer." *Cell* **128**(4): 683–692.

Jordan, G., and J. D. Mollon (1993). "A study of women heterozygous for colour deficiencies." *Vision Res* **33**(11): 1495–1508.

Jorgensen, A. L., J. Philip, et al. (1992). "Different patterns of X inactivation in MZ twins discordant for red-green color-vision deficiency." *Am J Hum Genet* **51**(2): 291–298.

Joyce, P. R., S. A. Williamson, et al. (2007). "Effects of childhood experiences on cortisol levels in depressed adults." *Aust N Z J Psychiatry* **41**(1): 62–65.

Junien, C., and P. Nathanielsz (2007). "Report on the IASO Stock Conference 2006: Early and lifelong environmental epigenomic programming of metabolic syndrome, obesity and type II diabetes." *Obes Rev* **8**(6): 487–502.

Just, E. E. (1939). *The biology of the cell surface.* Philadelphia: Blakison's.

Kaitz, M., H. R. Maytal, et al. (2010). "Maternal anxiety, mother-infant interactions, and infants' response to challenge." *Infant Behav Dev* **33**(2): 136–148.

Kaminsky, Z., A. Petronis, et al. (2008). "Epigenetics of personality traits: An illustrative study of identical twins discordant for risk-taking behavior." *Twin Res Hum Genet* **11**(1): 1–11.

Kapoor, A., A. Kostaki, et al. (2009). "The effects of prenatal stress on learning in adult offspring is dependent on the timing of the stressor." *Behav Brain Res* **197**(1): 144–149.

Kapoor, A., J. Leen, et al. (2008). "Molecular regulation of the hypothalamic-pituitary-adrenal axis in adult male guinea pigs after prenatal stress at different stages of gestation." *J Physiol* **586**(Pt 17): 4317–4326.

Kapoor, A., and S. G. Matthews (2008). "Prenatal stress modifies behavior and hypothalamic-pituitary-adrenal function in female guinea pig offspring: Effects of timing of prenatal stress and stage of reproductive cycle." *Endocrinology* **149**(12): 6406–6415.

Kapoor, A., S. Petropoulos, et al. (2008). "Fetal programming of hypothalamic-pituitary-adrenal (HPA) axis function and behavior by synthetic glucocorticoids." *Brain Res Rev* **57**(2): 586–595.

Kato, T. (2009). "Epigenomics in psychiatry." *Neuropsychobiology* **60**(1): 2–4.

Kato, T., K. Iwamoto, et al. (2005). "Genetic or epigenetic difference causing discordance between monozygotic twins as a clue to molecular basis of mental disorders." *Mol Psychiatry* **10**(7): 622–630.

Katz, L. A. (2006). "Genomes: Epigenomics and the future of genome sciences." *Curr Biol* **16**(23): R996–R997.

Kenny, P. A., and M. J. Bissell (2003). "Tumor reversion: Correction of malignant behavior by microenvironmental cues." *Int J Cancer* **107**(5): 688–695.

Kenny, P. A., G. Y. Lee, et al. (2007). "Targeting the tumor microenvironment." *Front Biosci* **12**: 3468–3474.

Kim, J. B., H. Zaehres, et al. (2008). "Pluripotent stem cells induced from adult neural stem cells by reprogramming with two factors." *Nature* **454**(7204): 646–650.

Kim, K. C., S. Friso, et al. (2009). "DNA methylation, an epigenetic mechanism connecting folate to healthy embryonic development and aging." *J Nutr Biochem* **20**(12): 917–926.

Kimball, J. W. (2010). *Kimball's biology pages.* http://users.rcn.com/jkimball.ma.ultranet/BiologyPages/.

King, N. E., and J. D. Mellen (1994). "The effects of early experience on adult copulatory behavior in zoo-born chimpanzees (*Pan troglodytes*)." *Zoo Biol* **13**: 51–59.

Kinoshita, T., Y. Ikeda, et al. (2008). "Genomic imprinting: A balance between antagonistic roles of parental chromosomes." *Semin Cell Dev Biol* **19**(6): 574–579.

Kirschner, M., J. Gerhart, et al. (2000). "Molecular 'vitalism.'" *Cell* **100**(1): 79–88.

Ko, J. M., J. M. Kim, et al. (2010). "Influence of parental origin of the X chromosome on physical phenotypes and GH responsiveness of patients with Turner syndrome." *Clin Endocrinol (Oxf)* **73**(1): 66–71.

Kochanska, G., R. A. Barry, et al. (2009). "Early attachment organization moderates the parent-child mutually coercive pathway to children's antisocial conduct." *Child Dev* **80**(4): 1288–1300.

Koornneef, M., C. J. Hanhart, et al. (1991). "A genetic and physiological analysis of late flowering mutants in *Arabidopsis thaliana*." *Mol Gen Genet* **229**(1): 57–66.

Kraemer, S. (2000). "The fragile male." *BMJ* **321**(7276): 1609–1612.

Kulesa, P. M., J. C. Kasemeier-Kulesa, et al. (2006). "Reprogramming metastatic melanoma cells to assume a neural crest cell-like phenotype in an embryonic microenvironment." *Proc Natl Acad Sci USA* **103**(10): 3752–3757.

Lanctot, C., T. Cheutin, et al. (2007). "Dynamic genome architecture in the nuclear space: Regulation of gene expression in three dimensions." *Nat Rev Genet* **8**(2): 104–115.

Lander, E. S., L. M. Linton, et al. (2001). "Initial sequencing and analysis of the human genome." *Nature* **409**(6822): 860–921.

Lanza, R. P., J. B. Cibelli, et al. (2000). "Cloning of an endangered species (*Bos gaurus*) using interspecies nuclear transfer." *Cloning* **2**(2): 79–90.

Laprise, S. L. (2009). "Implications of epigenetics and genomic imprinting in assisted reproductive technologies." *Mol Reprod Dev* **76**(11): 1006–1018.

Laugharne, J., A. Janca, et al. (2007). "Posttraumatic stress disorder and terrorism: 5 years after 9/11." *Curr Opin Psychiatry* **20**(1): 36–41.

Leeming, R. J., and M. Lucock (2009). "Autism: Is there a folate connection?" *J Inherit Metab Dis* **32**(3): 400–402.

Lewis, A., and W. Reik (2006). "How imprinting centres work." *Cytogenet Genome Res* **113**(1–4): 81–89.

Li, Y., Y. Dai, et al. (2006). "Cloned endangered species takin (*Budorcas taxicolor*) by inter-species nuclear transfer and comparison of the blastocyst development with yak (*Bos grunniens*) and bovine." *Mol Reprod Dev* **73**(2): 189–195.

Li, Y., Y. Dai, et al. (2007). "In vitro development of yak (*Bos grunniens*) embryos generated by interspecies nuclear transfer." *Anim Reprod Sci* **101**(1–2): 45–59.

Lightman, S. L. (2008). "The neuroendocrinology of stress: A never ending story." *J Neuroendocrinol* **20**(6): 880–884.

Lillycrop, K. A., E. S. Phillips, et al. (2005). "Dietary protein restriction of pregnant rats induces and folic acid supplementation prevents epigenetic modification of hepatic gene expression in the offspring." *J Nutr* **135**(6): 1382–1386.

Lillycrop, K. A., J. L. Slater-Jefferies, et al. (2007). "Induction of altered epigenetic regulation of the hepatic glucocorticoid receptor in the offspring of rats fed a protein-restricted diet during pregnancy suggests that reduced DNA methyltransferase-1 expression is involved in impaired DNA methylation and changes in histone modifications." *Br J Nutr* **97**(6): 1064–1073.

Liu, D., J. Diorio, et al. (1997). "Maternal care, hippocampal glucocorticoid receptors, and hypothalamic-pituitary-adrenal responses to stress." *Science* **277**(5332): 1659–1662.

Loat, C. S., K. Asbury, et al. (2004). "X inactivation as a source of behavioural differences in monozygotic female twins." *Twin Res* **7**(1): 54–61.

Loh, R., D. Hayes, et al. (2006). "The immunohistochemical characterization of devil facial tumor disease (DFTD) in the Tasmanian Devil (*Sarcophilus harrisii*)." *Vet Pathol* **43**(6): 896–903.

Lorthongpanich, C., C. Laowtammathron, et al. (2008). "Development of interspecies cloned monkey embryos reconstructed with bovine enucleated oocytes." *J Reprod Dev* **54**(5): 306–313.

Lotem, J., and L. Sachs (2002). "Epigenetics wins over genetics: Induction of differentiation in tumor cells." *Semin Cancer Biol* **12**(5): 339–346.

Lu, J., G. Getz, et al. (2005). "MicroRNA expression profiles classify human cancers." *Nature* **435**(7043): 834–838.

Lumey, L. H. (1998). "Reproductive outcomes in women prenatally exposed to undernutrition: A review of findings from the Dutch famine birth cohort." *Proc Nutr Soc* **57**(1): 129–135.

Lumey, L. H., and A. D. Stein (1997). "In utero exposure to famine and subsequent fertility: The Dutch Famine Birth Cohort Study." *Am J Public Health* **87**(12): 1962–1966.

Lyon, M. F. (1961). "Gene action in the X-chromosome of the mouse (*Mus musculus* L.)." *Nature* **190**(4773): 372–373.

Lyon, M. F. (1971). "Possible mechanisms of X chromosome inactivation." *Nat New Biol* **232**(34): 229–232.

Lyon, M. F. (1989). "X-chromosome inactivation as a system of gene dosage compensation to regulate gene expression." *Prog Nucleic Acid Res Mol Biol* **36**: 119–130.

Lyon, M. F. (1995). "The history of X-chromosome inactivation and relation of recent findings to understanding of human X-linked conditions." *Turk J Pediatr* **37**(2): 125–140.

Lyssiotis, C. A., R. K. Foreman, et al. (2009). "Reprogramming of murine fibroblasts to induced pluripotent stem cells with chemical complementation of Klf4." *Proc Natl Acad Sci USA* **106**(22): 8912–8917.

Maestripieri, D. (2003). "Similarities in affiliation and aggression between cross-fostered rhesus macaque females and their biological mothers." *Dev Psychobiol* **43**(4): 321–327.

Maestripieri, D. (2005). "Early experience affects the intergenerational transmission of infant abuse in rhesus monkeys." *Proc Natl Acad Sci USA* **102**(27): 9726–9729.

Maienschein, J. (2008). "Epigenesis and preformationism." *Stanford encyclopedia of philosophy*, ed. E. N. Zalta. Stanford, CA: Stanford University.

Martin, D. I., J. E. Cropley, et al. (2008). "Environmental influence on epigenetic inheritance at the *Avy* allele." *Nutr Rev* **66**(Suppl 1): S12–S14.

Martin, D. I., R. Ward, et al. (2005). "Germline epimutation: A basis for epigenetic disease in humans." *Ann N Y Acad Sci* **1054**: 68–77.

Masip, M., A. Veiga, et al. (2010). "Reprogramming with defined factors:

From induced pluripotency to induced transdifferentiation." *Mol Hum Reprod* **16**(11): 856–868.

Mastroeni, D., A. McKee, et al. (2009). "Epigenetic differences in cortical neurons from a pair of monozygotic twins discordant for Alzheimer's disease." *PLoS One* **4**(8): e6617.

Mattick, J. (2003). "Challenging the dogma: The hidden layer of non-protein-coding RNAs in complex organisms." *BioEssays* **25**: 930–939.

Mattick, J. S., and I. Makunin (2006). "Non-coding RNA." *Hum Mol Genet* **15**: R17–R29.

McCallum, H. (2008). "Tasmanian devil facial tumour disease: Lessons for conservation biology." *Trends Ecol Evol* **23**(11): 631–637.

McClellan, J., and M. C. King (2010). "Genetic heterogeneity in human disease." *Cell* **141**(2): 210–217.

McCormack, K., M. M. Sanchez, et al. (2006). "Maternal care patterns and behavioral development of rhesus macaque abused infants in the first 6 months of life." *Dev Psychobiol* **48**(7): 537–550.

McGowan, P. O., A. Sasaki, et al. (2009). "Epigenetic regulation of the glucocorticoid receptor in human brain associates with childhood abuse." *Nat Neurosci* **12**(3): 342–348.

Mcmillen, I. C., and J. S. Robinson (2005). "Developmental origins of the metabolic syndrome: Prediction, plasticity, and programming." *Physiol Rev* **85**(2): 571–633.

Meaney, M. J., M. Szyf, et al. (2007). "Epigenetic mechanisms of perinatal programming of hypothalamic-pituitary-adrenal function and health." *Trends Mol Med* **13**(7): 269–277.

Meder, A. (1993). "The effect of familiarity, age, dominance and rearing on reproductive success of captive gorillas." In *International studbook for the gorilla*, ed. R. Kirchshofer, 227–236. Frankfurt, Frankfurt Zoological Garden.

Michaud, E. J., M. J. van Vugt, et al. (1994). "Differential expression of a new dominant agouti allele (Aiapy) is correlated with methylation state and is influenced by parental lineage." *Genes Dev* **8**(12): 1463–1472.

Milnes, M. R., D. S. Bermudez, et al. (2006). "Contaminant-induced feminization and demasculinization of nonmammalian vertebrate males in aquatic environments." *Environ Res* **100**(1): 3–17.

Miltenberger, R. J., R. L. Mynatt, et al. (1997). "The role of the agouti gene in the yellow obese syndrome." *J Nutr* **127**(9): 1902S–1907S.

Mintz, B., and K. Illmensee (1975). "Normal genetically mosaic mice produced from malignant teratocarcinoma cells." *Proc Natl Acad Sci USA* **72**(9): 3585–3589.

Monroy, A. (1986). "A centennial debt of developmental biology to the sea urchin." *Biol Bull* **171**: 509–519.

Morak, M., H. K. Schackert, et al. (2008). "Further evidence for heritability of an epimutation in one of 12 cases with MLH1 promoter methylation in blood cells clinically displaying HNPCC." *Eur J Hum Genet* **16**(7): 804–811.

Moreira de Mello, J. C., E. S. de Araujo, et al. (2010). "Random X inactivation and extensive mosaicism in human placenta revealed by analysis of allele-specific gene expression along the X chromosome." *PLoS One* **5**(6): e10947.

Morgan, H., H. G. Sutherland, et al. (1999). "Epigenetic inheritance at the agouti locus in the mouse." *Nat Genet* **23**: 314–318.

Moss, L. (1992). "A kernel of truth? On the reality of the genetic program." *Phil Sci Assoc Proc* **1992**: 335–348.

Murchison, E. P., C. Tovar, et al. (2010). "The Tasmanian devil transcriptome reveals Schwann cell origins of a clonally transmissible cancer." *Science* **327**(5961): 84–87.

Murgia, C., J. K. Pritchard, et al. (2006). "Clonal origin and evolution of a transmissible cancer." *Cell* **126**(3): 477–487.

Murphy, S. K., and R. L. Jirtle (2003). "Imprinting evolution and the price of silence." *BioEssays* **25**(6): 577–588.

Namekawa, S. H., J. L. VandeBerg, et al. (2007). "Sex chromosome silencing in the marsupial male germ line." *Proc Natl Acad Sci USA* **104**(23): 9730–9735.

Nathans, J. (1999). "The evolution and physiology of human color vision: Insights from molecular genetic studies of visual pigments." *Neuron* **24**(2): 299–312.

Nathans, J., T. P. Piantanida, et al. (1986). "Molecular genetics of inherited variation in human color vision." *Science* **232**(4747): 203–210.

Nathans, J., D. Thomas, et al. (1986). "Molecular genetics of human color vision: The genes encoding blue, green, and red pigments." *Science* **232**(4747): 193–202.

Neel, J. V. (1962). "Diabetes mellitus: A 'thrifty' genotype rendered detrimental by 'progress'?" *Am J Hum Genet* **14**: 353–362.

Neel, J. V. (1999). "The 'thrifty genotype' in 1998." *Nutr Rev* **57**(5 Pt 2): S2–9.

Nelson, C. M., and M. J. Bissell (2006). "Of extracellular matrix, scaffolds, and signaling: Tissue architecture regulates development, homeostasis, and cancer." *Annu Rev Cell Dev Biol* **22**: 287–309.

Neugebauer, R., H. W. Hoek, et al. (1999). "Prenatal exposure to wartime famine and development of antisocial personality disorder in early adulthood." *JAMA* **282**(5): 455–462.

Newman, S. A. (2009). "E. E. Just's 'independent irritability' revisited: The activated egg as excitable soft matter." *Mol Reprod Dev* **76**(10): 966–974.

Nicholson, J. M., and P. Duesberg (2009). "On the karyotypic origin and evolution of cancer cells." *Cancer Genet Cytogenet* **194**(2): 96–110.

Niemann, H., X. C. Tian, et al. (2008). "Epigenetic reprogramming in embryonic and foetal development upon somatic cell nuclear transfer cloning." *Reproduction* **135**(2): 151–163.

Nijhout, H. F. (1990). "Metaphors and the role of genes in development." *BioEssays* **12**(9): 441–446.

Nobrega, M. A., Y. Zhu, et al. (2004). "Megabase deletions of gene deserts result in viable mice." *Nature* **431**(7011): 988–993.

O'Brien S. J., D. E. Wildt, et al. (1983). "The cheetah is depauperate in genetic variation." *Science* **221**(4609): 459–462.

Ohno, S. (1969). "The preferential activation of maternally derived alleles in development of interspecific hybrids." *Wistar Inst Symp Monogr* **9**: 137–150.

Okano, H. (2009). "Strategies toward CNS-regeneration using induced pluripotent stem cells." *Genome Inform* **23**(1): 217–220.

Otani, K., A. Suzuki, et al. (2009). "Effects of the 'affectionless control' parenting style on personality traits in healthy subjects." *Psychiatry Res* **165**(1–2): 181–186.

Ou, L., X. Wang, et al. (2010). "Is iPS cell the panacea?" *IUBMB Life* **62**(3): 170–175.

Owen, C. M., and J. H. Segars, Jr. (2009). "Imprinting disorders and assisted reproductive technology." *Semin Reprod Med* **27**(5): 417–428.

Oyama, S. (1985). *The ontogeny of information: Developmental systems and evolution.* Cambridge, UK: Cambridge University Press.

Painter, R. C., C. Osmond, et al. (2008). "Transgenerational effects of prenatal exposure to the Dutch famine on neonatal adiposity and health in later life." *BJOG* **115**(10): 1243–1249.

Painter, R. C., T. J. Roseboom, et al. (2005). "Adult mortality at age 57 after prenatal exposure to the Dutch famine." *Eur J Epedemiol* **20**(8): 673–676.

Pardo, P. J., A. L. Pérez, et al. (2007). "An example of sex-linked color vision differences." *Color Res Appl* **32**(6): 433–439.

Parikh, V. N., T. Clement, et al. (2006). "Physiological consequences of social descent: Studies in *Astatotilapia burtoni.*" *J Endocrinol* **190**(1): 183–190.

Passier, R., and C. Mummery (2003). "Origin and use of embryonic and adult stem cells in differentiation and tissue repair." *Cardiovasc Res* **58**(2): 324–335.

Pathak, S., and A. S. Multani (2006). "Aneuploidy, stem cells and cancer." *EXS* **96**: 49–64.

Patton, G. C., C. Coffey, et al. (2001). "Parental 'affectionless control' in adolescent depressive disorder." *Soc Psychiatry Psychiatr Epidemiol* **36**(10): 475–480.

Pearse, A. M., and K. Swift (2006). "Allograft theory: Transmission of devil facial-tumour disease." *Nature* **439**(7076): 549.

Pembrey, M. E., L. O. Bygren, et al. (2006). "Sex-specific, male-line transgenerational responses in humans." *Eur J Hum Genet* **14**(2): 159–166.

Petronis, A. (2001). "Human morbid genetics revisited: relevance of epigenetics." *Trends Genet* **17**(3): 142–146.

Pezer, Z., and D. Ugarkovic (2008). "Role of non-coding RNA and heterochromatin in aneuploidy and cancer." *Semin Cancer Biol* **18**(2): 123–130.

Pigliucci, M. (2010). "Genotype-phenotype mapping and the end of the 'genes as blueprint' metaphor." *Phil Trans Royal Soc B* **365**(1540): 557–566.

Pollard, K. S., S. R. Salama, et al. (2006). "An RNA gene expressed during cortical development evolved rapidly in humans." *Nature* **443**(7108): 167–172.

Popova, B. C., T. Tada, et al. (2006). "Attenuated spread of X-inactivation in an X-autosome translocation." *Proc Natl Acad Sci USA* **103**(20): 7706–7711.

Portin, P. (2009). "The elusive concept of the gene." *Hereditas* **146**(3): 112–117.

Porton, I., and K. Niebrugge (2002). The changing role of handrearing in zoo-based primate breeding programs. In *Developments in primatology, Progress and prospects: Nursery rearing of nonhuman primates in the 21st century*, ed. G. P. Sackett, G. C. Ruppenthal, and K. Elias, 21–31. New York: Springer.

Prentice, A. M., B. J. Hennig, et al. (2008). "Evolutionary origins of the obesity epidemic: Natural selection of thrifty genes or genetic drift following predation release?" *Int J Obes (Lond)* **32**(11): 1607–1610.

Prins, G. S. (2008). "Endocrine disruptors and prostate cancer risk." *Endocr Relat Cancer* **15**(3): 649–656.

Provine, W. B. (1986). *Sewall Wright and evolutionary biology*. Cambridge, MA: MIT Press.

Ptak, C., and A. Petronis (2010). "Epigenetic approaches to psychiatric disorders." *Dialogues Clin Neurosci* **12**(1): 25–35.

Puri, D., J. Dhawan, et al. (2010). "The paternal hidden agenda: Epigenetic inheritance through sperm chromatin." *Epigenetics* **5**(5).

Rakyan, V., M. Blewitt, et al. (2002). "Metastable epialleles in mammals." *Trends Genet* **18**: 348–353.

Rakyan, V. K., S. Chong, et al. (2003). "Transgenerational inheritance of epigenetic states at the murine *Axin(Fu)* allele occurs after maternal and paternal transmission." *Proc Natl Acad Sci USA* **100**(5): 2538–2543.

Rakyan, V. K., J. Preis, et al. (2001). "The marks, mechanisms and memory of epigenetic states in mammals." *Biochem J* **356**(Pt 1): 1–10.

Rassoulzadegan, M., V. Grandjean, et al. (2006). "RNA-mediated non-Mendelian inheritance of an epigenetic change in the mouse." *Nature* **441**(7092): 469–474.

Rassoulzadegan, M., V. Grandjean, et al. (2007). "Inheritance of an epigenetic change in the mouse: A new role for RNA." *Biochem Soc Trans* **35**(3): 623–625.

Ravelli, A. C., J. H. van der Meulen, et al. (1998). "Glucose tolerance in adults after prenatal exposure to famine." *Lancet* **351**(9097): 173–177.

Ravelli, G. P., Z. A. Stein, et al. (1976). "Obesity in young men after famine exposure in utero and early infancy." *N Engl J Med* **295**(7): 349–353.

Reik, W. (1989). "Genomic imprinting and genetic disorders in man." *Trends Genet* **5**(10): 331–336.

Reik, W., M. Constancia, et al. (2003). "Regulation of supply and demand for maternal nutrients in mammals by imprinted genes." *J Physiol* **547**(Pt 1): 35–44.

Reik, W., W. Dean, et al. (2001). "Epigenetic reprogramming in mammalian development." *Science* **293**: 1089–1092.

Renn, S. C., N. Aubin-Horth, et al. (2008). "Fish and chips: Functional genomics of social plasticity in an African cichlid fish." *J Exp Biol* **211**(Pt 18): 3041–3056.

Revollo, J. R., and J. A. Cidlowski (2009). "Mechanisms generating diversity in glucocorticoid receptor signaling." *Ann N Y Acad Sci* **1179**: 167–178.

Rheinberger, H.-J. (2008). "Gene." *Stanford encyclopedia of philosophy.* Stanford, CA: Stanford University.

Richards, E. J. (2006). "Inherited epigenetic variation—revisiting soft inheritance." *Nat Rev Genet* **7**(5): 395–401.

Riggs, A. D. (2002). "X chromosome inactivation, differentiation, and DNA methylation revisited, with a tribute to Susumu Ohno." *Cytogenet Genome Res* **99**(1–4): 17–24.

Rodriguez-Carmona, M., L. T. Sharpe, et al. (2008). "Sex-related differences in chromatic sensitivity." *Vis Neurosci* **25**(3): 433–440.

Roemer, I., W. Reik, et al. (1997). "Epigenetic inheritance in the mouse." *Curr Biol* **7**: 277–280.

Rogers, E. J. (2008). "Has enhanced folate status during pregnancy altered natural selection and possibly autism prevalence? A closer look at a possible link." *Med Hypotheses* **71**(3): 406–410.

Roseboom, T. J., S. de Rooij, et al. (2006). "The Dutch famine and its long-term consequences for adult health." *Early Hum Dev* **82**(8): 485–491.

Roseboom, T. J., J. H. van der Meulen, et al. (1999). "Blood pressure in adults after prenatal exposure to famine." *J Hypertens* **17**(3): 325–330.

Roseboom, T. J., J. H. van der Meulen, et al. (2000a). "Coronary heart disease after prenatal exposure to the Dutch famine, 1944–45." *Heart* **84**(6): 595–598.

Roseboom, T. J., J. H. van der Meulen, et al. (2000b). "Plasma lipid profiles in adults after prenatal exposure to the Dutch famine." *Am J Clin Nutr* **72**(5): 1101–1106.

Ross, H. E., and L. J. Young (2009). "Oxytocin and the neural mechanisms regulating social cognition and affiliative behavior." *Front Neuroendocrinol* **30**(4): 534–547.

Ross, J., D. Roeltgen, et al. (2006). "Cognition and the sex chromosomes: Studies in Turner syndrome." *Horm Res* **65**(1): 47–56.

Rothwell, N. J., and M. J. Stock (1981). "Regulation of energy balance." *Annu Rev Nutr* **1**: 235–256.

Ruppenthal, G. C., G. L. Arling, et al. (1976). "A 10-year perspective of motherless-mother monkey behavior." *J Abnorm Psychol* **85**(4): 341–349.

Ryan, S., S. Thompson, et al. (2002). "Effects of hand-rearing on the reproductive success of western lowland gorillas in North America." *Zoo Biol* **21**: 389–401.

Sales, K. M., M. C. Winslet, et al. (2007). "Stem cells and cancer: An overview." *Stem Cell Rev* **3**(4): 249–255.

Sanjayan, M. A., and K. Crooks (1996). "Skin grafts and cheetahs." *Nature* **381**(6583): 566.

Santos, F., and W. Dean (2004). "Epigenetic reprogramming during early development in mammals." *Reproduction* **127**(6): 643–651.

Sapp, J. (2009). " 'Just' in time: Gene theory and the biology of the cell surface." *Mol Reprod Dev* **76**(10): 903–911.

Schickel, R., B. Boyerinas, et al. (2008). "MicroRNAs: Key players in the immune system, differentiation, tumorigenesis and cell death." *Oncogene* **27**(45): 5959–5974.

Schier, A. F. (2007). "The maternal-zygotic transition: Death and birth of RNAs." *Science* **316**(5823): 406–407.

Schubeler, D. (2009). "Epigenomics: Methylation matters." *Nature* **462**(7271): 296–297.

Schulz, W. A., and J. Hatina (2006). "Epigenetics of prostate cancer: Beyond DNA methylation." *J Cell Mol Med* **10**(1): 100–125.

Schulze-Tanzil, G. (2009). "Activation and dedifferentiation of chondrocytes: Implications in cartilage injury and repair." *Ann Anat* **191**(4): 325–338.

Schwanzel-Fukuda, M., K. L. Jorgenson, et al. (1992). "Biology of normal luteinizing hormone-releasing hormone neurons during and after their migration from olfactory placode." *Endocr Rev* **13**(4): 623–634.

Seckl, J. R. (2004). "Prenatal glucocorticoids and long-term programming." *Eur J Endocrinol* **151**(Suppl 3): U49–U62.

Seckl, J. R. (2008). "Glucocorticoids, developmental 'programming' and the risk of affective dysfunction." *Prog Brain Res* **167**: 17–34.

Seckl, J. R., and M. C. Holmes (2007). "Mechanisms of disease: Glucocorticoids, their placental metabolism and fetal 'programming' of adult pathophysiology." *Nat Clin Pract Endocrinol Metab* **3**(6): 479–488.

Seckl, J. R., and M. J. Meaney (2006). "Glucocorticoid 'programming' and PTSD risk." *Ann N Y Acad Sci* **1071**: 351–378.

Serbin, L. A., and J. Karp (2004). "The intergenerational transfer of psychosocial risk: Mediators of vulnerability and resilience." *Annu Rev Psychol* **55**: 333–363.

Shi, W., A. Krella, et al. (2005). "Widespread disruption of genomic imprinting in adult interspecies mouse (*Mus*) hybrids." *Genesis* **43**(3): 100–108.

Shire, J. G. (1989). "Unequal parental contributions: Genomic imprinting in mammals." *New Biol* **1**(2): 115–120.

Shively, C. A., T. C. Register, et al. (2009). "Social stress, visceral obesity, and coronary artery atherosclerosis: Product of a primate adaptation." *Am J Primatol* **71**(9): 742–751.

Shuldiner, A. R., and K. M. Munir (2003). "Genetics of obesity: More complicated than initially thought." *Lipids* **38**(2): 97–101.

Shyue, S. K., D. Hewett-Emmett, et al. (1995). "Adaptive evolution of color vision genes in higher primates." *Science* **269**(5228): 1265–1267.

Siddle, H. V., A. Kreiss, et al. (2007). "Transmission of a fatal clonal tumor by biting occurs due to depleted MHC diversity in a threatened carnivorous marsupial." *Proc Natl Acad Sci USA* **104**(41): 16221–16226.

Sikela, J. M. (2006). "The jewels of our genome: The search for the genomic changes underlying the evolutionarily unique capacities of the human brain." *PLoS Genet* **2**(5): e80.

Simmons, R. A. (2007). "Developmental origins of diabetes: The role of epigenetic mechanisms." *Curr Opin Endocrinol Diabetes Obes* **14**(1): 13–16.

Singh, S. M., and R. O'Reilly (2009). "(Epi)genomics and neurodevelopment in schizophrenia: Monozygotic twins discordant for schizophrenia augment the search for disease-related (epi)genomic alterations." *Genome* **52**(1): 8–19.

Skinner, M. K., M. Manikkam, et al. (2010). "Epigenetic transgenerational actions of environmental factors in disease etiology." *Trends Endocrinol Metab* **21**(4): 214–222.

Skuse, D. H., R. S. James, et al. (1997). "Evidence from Turner's syndrome of an imprinted X-linked locus affecting cognitive function." *Nature* **387**(6634): 705–708.

Smith, C. (1947). "The effects of wartime starvation in Holland on pregnancy and its product." *Am J Obst Gynecol* **53**: 599–608.

Smith, F. M., L. J. Holt, et al. (2007). "Mice with a disruption of the imprinted *Grb10* gene exhibit altered body composition, glucose homeostasis, and insulin signaling during postnatal life." *Mol Cell Biol* **27**(16): 5871–5886.

Smithies, O. (2005). "Many little things: one geneticist's view of complex diseases." *Nat Rev Genet* **6**(5): 419–425.

Snell, G. D., and S. Reed (1993). "William Ernest Castle, pioneer mammalian geneticist." *Genetics* **133**(4): 751–753.

Song, B. S., S. H. Lee, et al. (2009). "Nucleologenesis and embryonic genome activation are defective in interspecies cloned embryos between bovine ooplasm and rhesus monkey somatic cells." *BMC Dev Biol* **9**: 44.

Soto, A. M., and C. Sonnenschein (2004). "The somatic mutation theory of cancer: Growing problems with the paradigm?" *BioEssays* **26**(10): 1097–1107.

Soto, A. M., and C. Sonnenschein (2010). "Environmental causes of cancer: Endocrine disruptors as carcinogens." *Nat Rev Endocrinol* **6**(7): 363–370.

Speakman, J. R. (2006). "Thrifty genes for obesity and the metabolic syndrome—time to call off the search?" *Diab Vasc Dis Res* **3**(1): 7–11.

Speakman, J. R. (2008). "Thrifty genes for obesity, an attractive but flawed idea, and an alternative perspective: The 'drifty gene' hypothesis." *Int J Obes (Lond)* **32**(11): 1611–1617.

Stein, A. D., A. C. Ravelli, et al. (1995). "Famine, third-trimester pregnancy weight gain, and intrauterine growth: The Dutch Famine Birth Cohort Study." *Hum Biol* **67**(1): 135–150.

Stein, Z., and M. Susser (1975). "The Dutch famine, 1944–1945, and the reproductive process. II. Interrelations of caloric rations and six indices at birth." *Pediatr Res* **9**(2): 76–83.

Stein, Z., M. Susser, et al. (1972). "Nutrition and mental performance." *Science* **178**: 706–713.

Stocum, D. L. (2002). "Regenerative biology and medicine." *J Musculoskelet Neuronal Interact* **2**(3): 270–273.

Stocum, D. L. (2004). "Amphibian regeneration and stem cells." *Curr Top Microbiol Immunol* **280**: 1–70.

Stokes, T. L., B. N. Kunkel, et al. (2002). "Epigenetic variation in *Arabidopsis* disease resistance. *Genes Dev* **16**(2): 171–182.

Stokes, T. L., and E. J. Richards (2002). "Induced instability of two *Arabidopsis* constitutive pathogen-response alleles." *Proc Natl Acad Sci USA* **99**(11): 7792–7796.

Stoltz, K., P. E. Griffiths, et al. (2004). "How biologists conceptualize genes: An empirical study." *Stud Hist Phil Biol Biomed Sci* **35**: 647–673.

Storz, G., S. Altuvia, et al. (2005). "An abundance of RNA regulators." *Annu Rev Biochem* **74**: 199–217.

Stouder, C., and A. Paoloni-Giacobino (2010). "Transgenerational effects of the endocrine disruptor vinclozolin on the methylation pattern of imprinted genes in the mouse sperm." *Reproduction* **139**(2): 373–379.

Strahl, B. D., and C. D. Allis (2000). "The language of covalent histone modifications." *Nature* **403**(6765): 41–45.

Straube, W. L., and E. M. Tanaka (2006). "Reversibility of the differentiated state: Regeneration in amphibians." *Artif Organs* **30**(10): 743–755.

Suay, F., A. Salvador, et al. (1999). "Effects of competition and its outcome on serum testosterone, cortisol and prolactin." *Psychoneuroendocrinology* **24**(5): 551–566.

Suijkerbuijk, K. P., E. van der Wall, et al. (2007). "[Epigenetic processes in malignant transformation: The role of DNA methylation in cancer development]." *Ned Tijdschr Geneeskd* **151**(16): 907–913.

Sun, Y. H., S. P. Chen, et al. (2005). "Cytoplasmic impact on cross-genus cloned fish derived from transgenic common carp (*Cyprinus carpio*) nuclei and goldfish (*Carassius auratus*) enucleated eggs." *Biol Reprod* **72**(3): 510–515.

Susser, M., and B. Levin (1999). "Ordeals for the fetal programming hypothesis. The hypothesis largely survives one ordeal but not another." *BMJ* **318**(7188): 885–886.

Swarbrick, M. M., and C. Vaisse (2003). "Emerging trends in the search for genetic variants predisposing to human obesity." *Curr Opin Clin Nutr Metab Care* **6**(4): 369–375.

Szyf, M., I. C. Weaver, et al. (2005). "Maternal programming of steroid receptor expression and phenotype through DNA methylation in the rat." *Front Neuroendocrinol* **26**(3–4): 139–162.

Taft, R. J., M. Pheasant, et al. (2007). "The relationship between non-protein-coding DNA and eukaryotic complexity." *BioEssays* **29**(3): 288–299.

Takahashi, K., K. Okita, et al. (2007). "Induction of pluripotent stem cells from fibroblast cultures." *Nat Protoc* **2**(12): 3081–3089.

Taylor, P. D., and L. Poston (2007). "Developmental programming of obesity in mammals." *Exp Physiol* **92**(2): 287–298.

ten Berge, D., W. Koole, et al. (2009). "Wnt signaling mediates self-organization and axis formation in embryoid bodies." *Cell Stem Cell* **3**(5): 508–515.

Thomas, C. A. J. (1971). "The genetic organization of chromosomes." *Annu Rev Genet* **5**: 237–256.

Thongphakdee, A., S. Kobayashi, et al. (2008). "Interspecies nuclear transfer embryos reconstructed from cat somatic cells and bovine ooplasm." *J Reprod Dev* **54**(2): 142–147.

Tiberio, G. (1994). "MZ female twins discordant for X-linked diseases: A review." *Acta Genet Med Gemellol (Roma)* **43**(3–4): 207–214.

Tobi, E. W., L. H. Lumey, et al. (2009). "DNA methylation differences after exposure to prenatal famine are common and timing- and sex-specific." *Hum Mol Genet* **18**(21): 4046–4053.

Tokumoto, Y., S. Ogawa, et al. (2010). "Comparison of efficiency of terminal differentiation of oligodendrocytes from induced pluripotent stem cells versus embryonic stem cells in vitro." *J Biosci Bioeng* **109**(6): 622–628.

Tovee, M. J. (1993). "Colour vision in New World monkeys and the single-locus X-chromosome theory." *Brain Behav Evol* **42**(2): 116–127.

Trosko, J. E. (2009). "Review paper. Cancer stem cells and cancer non-stem cells: From adult stem cells or from reprogramming of differentiated somatic cells." *Vet Pathol Online* **46**(2): 176–193.

Tu, S. M., S. H. Lin, et al. (2002). "Stem-cell origin of metastasis and heterogeneity in solid tumours." *Lancet Oncol* **3**(8): 508–513.

Tweedell, K. (2008). New paths to pluripotent stem cells." *Curr Stem Cell Res Ther* **3**: 151–162.

Tyrka, A. R., L. Wier, et al. (2008). "Childhood parental loss and adult hypothalamic-pituitary-adrenal function." Biol. Psychiat. **63**(12): 1147–1154.

Uhm, S. J., M. K. Gupta, et al. (2007). "Expression of enhanced green fluorescent protein in porcine- and bovine-cloned embryos following interspecies somatic cell nuclear transfer of fibroblasts transfected by retrovirus vector." *Mol Reprod Dev* **74**: 1538–1547.

Urnov, F. D., and A. P. Wolffe (2001). "Above and within the genome: Epigenetics past and present." *J Mammary Gland Biol Neoplasia* **6**(2): 153–167.

VandeBerg, J. L., P. G. Johnston, et al. (1983). "X-chromosome inactivation and evolution in marsupials and other mammals." *Isozymes Curr Top Biol Med Res* **9**: 201–218.

Van Speybroeck, L., D. De Waele, et al. (2002). "Theories in early embryology." *Ann N Y Acad Sci* **981**: 7–49.

Ventolini, G., R. Neiger, et al. (2008). "Incidence of respiratory disorders in neonates born between 34 and 36 weeks of gestation following exposure to antenatal corticosteroids between 24 and 34 weeks of gestation." *Am J Perinatol* **25**(2): 79–83.

Verriest, G., and A. Gonella (1972). "An attempt at clinical determination by means of surface colours of the convergence points in congenital and acquired defects of colour vision." *Mod Probl Ophthalmol* **11**: 205–212.

Virtanen, H. E., E. Rajpert-De Meyts, et al. (2005). "Testicular dysgenesis syndrome and the development and occurrence of male reproductive disorders." *Toxicol Appl Pharmacol* **207**(2, Suppl): 501–505.

Voisey, J., and A. van Daal (2002). "Agouti: From mouse to man, from skin to fat." *Pigment Cell Res* **15**(1): 10–18.

Vos, J. G., E. Dybing, et al. (2000). "Health effects of endocrine-disrupting chemicals on wildlife, with special reference to the European situation." *Crit Rev Toxicol* **30**(1): 71–133.

Waddington, C. (1935/1946). *How animals develop.* London: George Allen & Unwin.

Waddington, C. (1962). *New patterns in genetics and development.* New York: Columbia University Press.

Waddington, C. (1968). "The basic ideas of biology." In *Towards a theoretical biology*, ed. C. Waddington. Vol. 1: *Prolegomena*, 1–32. Edinburgh: Edinburgh University Press.

Wadhwa, P. D., C. Buss, et al. (2009). "Developmental origins of health and disease: Brief history of the approach and current focus on epigenetic mechanisms." *Semin Reprod Med* **27**(5): 358–368.

Wagschal, A., and R. Feil (2006). "Genomic imprinting in the placenta." *Cytogenet Genome Res* **113**(1–4): 90–98.

Walker, B. R., and R. Andrew (2006). "Tissue production of cortisol by 11β-hydroxysteroid dehydrogenase type 1 and metabolic disease." *Ann N Y Acad Sci* **1083**: 165–184.

Warner, M. J., and S. E. Ozanne (2010). "Mechanisms involved in the developmental programming of adulthood disease." *Biochem J* **427**(3): 333–347.

Waterland, R. A., and K. B. Michels (2007). "Epigenetic epidemiology of the developmental origins hypothesis." *Annu Rev Nutr* **27**(1): 363–388.

Waterland, R. A., M. Travisano, et al. (2007). "Diet-induced hypermethylation at agouti viable yellow is not inherited transgenerationally through the female." *FASEB J* **21**(12): 3380–3385.

Watson, J. D. (1968). *The double helix: A personal account of the discovery of the structure of DNA.* New York: Atheneum.

Watson, J. D., and F. H. Crick (1953a). "Genetical implications of the structure of deoxyribonucleic acid." *Nature* **171**(4361): 964–967.

Watson, J. D., and F. H. Crick (1953b). "Molecular structure of nucleic acids; a structure for deoxyribose nucleic acid." *Nature* **171**(4356): 737–738.

Weaver, A., R. Richardson, et al. (2004). "Response to social challenge in young bonnet (*Macaca radiata*) and pigtail (*Macaca nemestrina*) macaques is related to early maternal experiences." *Am J Primatol* **62**(4): 243–259.

Weaver, I. C. (2009). "Shaping adult phenotypes through early life environments." *Birth Defects Res C Embryo Today* **87**(4): 314–326.

Weaver, I. C., N. Cervoni, et al. (2004). "Epigenetic programming by maternal behavior." *Nat Neurosci* **7**(8): 847–854.

Weaver, I. C., F. A. Champagne, et al. (2005). "Reversal of maternal programming of stress responses in adult offspring through methyl supplementation: Altering epigenetic marking later in life." *J Neurosci* **25**(47): 11045–11054.

Weaver, I. C., M. J. Meaney, et al. (2006). "Maternal care effects on the hippocampal transcriptome and anxiety-mediated behaviors in the offspring that are reversible in adulthood." *Proc Natl Acad Sci USA* **103**(9): 3480–3485.

Weksberg, R., C. Shuman, et al. (2005). "Beckwith-Wiedemann syndrome." *Am J Med Genet C Semin Med Genet* **137C**(1): 12–23.

Weksberg, R., and J. A. Squire (1996). "Molecular biology of Beckwith-Wiedemann syndrome." *Med Pediatr Oncol* **27**(5): 462–469.

Wells, J. C. (2009). "Ethnic variability in adiposity and cardiovascular risk: The variable disease selection hypothesis." *Int J Epidemiol* **38**(1): 63–71.

White, S. A., T. Nguyen, et al. (2002). "Social regulation of gonadotropin-releasing hormone." *J Exp Biol* **205**(Pt 17): 2567–2581.

Whitlock, K. E., N. Illing, et al. (2006). "Development of GnRH cells: Setting the stage for puberty." *Mol Cell Endocrinol* **254–255**: 39–50.

Williams, C. A., H. Angelman, et al. (1995). "Angelman syndrome: Consensus for diagnostic criteria. Angelman Syndrome Foundation." *Am J Med Genet* **56**(2): 237–238.

Wilmut, I., A. E. Schnieke, et al. (1997). "Viable offspring derived from fetal and adult mammalian cells." *Nature* **385**(6619): 810–813.

Wilson, B. D., M. M. Ollmann, et al. (1995). "Structure and function of ASP, the human homolog of the mouse agouti locus." *Hum Mol Genet* **4**(2): 223–230.

Witchel, S. F., and D. B. DeFranco (2006). "Mechanisms of disease: Regulation of glucocorticoid and receptor levels—impact on the metabolic syndrome." *Nat Clin Pract Endocrinol Metab* **2**(11): 621–631.

Wohlfahrt-Veje, C., K. M. Main, et al. (2009). "Testicular dysgenesis syndrome: Foetal origin of adult reproductive problems." *Clin Endocrinol (Oxf)* **71**(4): 459–465.

Wolff, G. L. (1996). "Variability in gene expression and tumor forma-

tion within genetically homogeneous animal populations in bioassays."
Fundam Appl Toxicol **29**(2): 176–184.

Wolff, G. L., R. L. Kodell, et al. (1998). "Maternal epigenetics and methyl supplements affect agouti gene expression in Avy/a mice." *FASEB J* **12**(11): 949–957.

Wolff, G. L., D. W. Roberts, et al. (1986). "Prenatal determination of obesity, tumor susceptibility, and coat color pattern in viable yellow (Avy/a) mice. The yellow mouse syndrome." *J Hered* **77**(3): 151–158.

Wolfram, S. (2002). *A new kind of science.* Champagn, IL: Wolfram Media.

Wong, A. H., Gottesman, I. I., et al. (2005). "Phenotypic differences in genetically identical organisms: The epigenetic perspective." *Hum Mol Genet* **14**(Spec 1): R11–R18.

Wright, S. (1916). "An intensive study of the inheritance of color and other coat characters in guinea pigs with special reference to graded variation." *Carnegie Inst Wash Publ* **241**: 59–160.

Wright, S. (1927). "The effects in combination of the major color factors of the guinea pig." *Genetics* **12**: 530–569.

Wroe, S., C. McHenry, et al. (2005). "Bite club: Comparative bite force in big biting mammals and the prediction of predatory behaviour in fossil taxa." *Proc Biol Sci* **272**(1563): 619–625.

Yan, S. Y., M. Tu, et al. (1990). "Developmental incompatibility between cell nucleus and cytoplasm as revealed by nuclear transplantation experiments in teleost of different families and orders." *Int J Dev Biol* **34**(2): 255–266.

Yehuda, R., A. Bell, et al. (2008). "Maternal, not paternal, PTSD is related to increased risk for PTSD in offspring of Holocaust survivors." *J Psychiatr Res* **42**(13): 1104–1111.

Yehuda, R., and L. M. Bierer (2007). "Transgenerational transmission of cortisol and PTSD risk." *Prog Brain Res* **167**: 121–135.

Yehuda, R., S. M. Engel, et al. (2005). "Transgenerational effects of post-traumatic stress disorder in babies of mothers exposed to the World Trade Center attacks during pregnancy." *J Clin Endocrinol Metab* **90**(7): 4115–4118.

Ying, S. Y., D. C. Chang, et al. (2008). "The microRNA (miRNA): Overview of the RNA genes that modulate gene function." *Mol Biotechnol* **38**(3): 257–268.

Youngson, N. A., and E. Whitelaw (2008). "Transgenerational epigenetic effects." *Annu Rev Genomics Hum Genet* **9**(1): 233–257.

Zeisel, S. H. (2009). "Importance of methyl donors during reproduction." *Am J Clin Nutr* **89**(2): 673S–677S.

Zilberman, D., and S. Henikoff (2005). "Epigenetic inheritance in *Arabidopsis*: Selective silence." *Curr Opin Genet Dev* **15**(5): 557–562.

Index

Page numbers in *italics* refer to illustrations.